DEWEY

Animals
That
Frighten
People

Animals That

by Dorothy E. Shuttlesworth

Frighten People
Fact versus Myth

illustrated with photographs

E. P. DUTTON & CO., INC.   NEW YORK

For Eddy Butler,
a remarkably handsome feline,
and for his "folks,"
Shirley and Earl

LIBRARY OF CONGRESS CATALOGING IN PUBLICATION DATA

Shuttlesworth, Dorothy Edwards, 1907–
Animals that frighten people.

1. Animals, Habits and behavior of—Juvenile literature.
2. Animal lore—Juvenile literature.   I. Title.
QL751.5.S55   596'.06'5   73-77458
ISBN 0-525-25745-4

Published simultaneously in Canada by Clarke,
Irwin & Company Limited, Toronto and Vancouver

Designed by Andrew Ross
Printed in the U.S.A.
First Edition

Contents

Introduction   *1*

1   The Big, Not-So-Bad Wolf   *5*

2   The Great Cats   *18*

3   Bear Hugs—Fatal or Friendly?   *30*

4   Gorilla—The Perfect Monster   *41*

5   Dracula Was Not a Bat   *51*

6   Snake Fright and Snakebite   *62*

7   Survivors from the Age of Reptiles:
    Alligators and Crocodiles   *80*

8   Terrors of the Sea:
    Sharks, Octopi, and Squids   *89*

9   Birds of Prey   *102*

10 Tarantulas, Other Spiders,
    and Scorpions   *110*

Index   *120*

Introduction

Today giant dinosaurs are among the most popular of animals. People delight in reading about them and having them recreated by scientists and artists. Many boys and girls say they "love" *Tyrannosaurus rex*—a huge, flesh-eating beast.

But there is no danger of anyone being gobbled up by *Tyrannosaurus*. It vanished from the earth, as did all dinosaurs, millions of years ago. After that, many other kinds of animals developed, and finally a variety of them began to share the world with people.

Until weapons were invented, all the big animals must have been frightening to humans. A number of the four-footed creatures were powerful, and were equipped with dagger-like teeth and sharp claws. They seemed better fitted for survival than people did.

Then gradually the tables were turned as men

developed skills that put animals at their mercy.
They learned how to produce fire and to make clubs,
bows and arrows, and firearms. They took over forests and grasslands, destroying the natural habitats
of wildlife.

In the course of these changes, certain animals
were domesticated and a few kinds became friendly
pets. But along with such peaceful associations,
there were hostile relationships, and a number of
animals continued to inspire fear. When farmers
lost their livestock to predators, or when ranchers'
herds were attacked, they were furious at the animals
that did the damage.

As a result, any creatures that proved a nuisance
in our modern world were considered "bad," as well
as frightening. "Good" animals, usually plant-eaters,
were those that caused no trouble. Little thought
was given to the fact that all kinds were simply
trying to survive by using the equipment and instincts which nature had provided.

In recent years people have become more understanding. They realize the earth needs a balance of
many kinds of life, and some efforts are being made
to correct the harm caused by centuries of destruc-
tive behavior. But often this is impossible. And in
spite of the concern and better understanding of animals, a variety of falsehoods are still believed about
them—fictions that were begun long ago and
passed on from one generation to another.

Of course we must acknowledge there are a few

features that understandably inspire fear: A gorilla's size and strength! A shark's knife-like teeth! The venom of some snakes and spiders! The sharp claws of the big cats! And, not surprisingly, we find the creatures that can inflict injury or death of special interest.

Animals That Frighten People, therefore, presents those that are noted for their dangerous potentials and evil reputations. The book also estimates the actual harm they can cause and the reasons for their actions, and considers how best to avoid trouble with them.

1

The Big,
Not-So-Bad Wolf

"Little Red Riding Hood" probably has done more than any other story to give people the impression that wolves are villains. It depicts this animal not only as a people-eater, but as a sly and tricky character.

Centuries ago, such a tale was typical of fables told in Europe. Supposedly, wolves were thoroughly evil; given half a chance, they would attack and kill humans—not so much because they were hungry as because they were vicious.

Such slander against these beasts did not end with ordinary kinds of accusation. Much folklore, based on superstitions, concerned werewolves ("man-wolves").

The werewolf stories varied. Humans might turn into wolves, or the other way around. According to the myths, people were transformed either by black magic or because they themselves wished to devour human flesh, and thus managed to assume a wolf-body in order to carry out their lust.

Today werewolf stories are still written, even dramatized in motion pictures. But it is understood that they are fantasy, designed for those who like a touch of the supernatural with their entertainment.

However, several hundred years ago, particularly in France, werewolves were taken seriously. Records show that every now and then a person was accused of being a werewolf and, assuming the animal's form, of devouring people. Sometimes the accused "werewolves" were brought to trial, found guilty, and executed.

The term used for the supposed crime was lycanthropy—a word coming from the ancient Greeks. The word werewolf was coined in England.

One startling fact, besides the fanciful stories about werewolves, is that there *have* been humans who actually believed a wolf's spirit had taken possession of them. They would howl, bite people, and run about on all fours. Doctors recognize this as a unique form of insanity.

In several ancient languages of Europe and Asia, wolves were called *leopes,* a name which sheds an interesting light on the wolf's relationship to other mammals. *Leo* means "lion"; *pes,* "footed." Yet "lion-footed" is not a true description of modern wolves. It does fit an extinct animal known as the dire wolf or cave wolf. The lion, being of the cat (feline) family, has sharp, retractile claws. The wolf,

Painting by Charles R. Knight,
Courtesy of the American Museum of Natural History

Wolves that lived before history was written are known to us through the discoveries of their fossil bones. These prehistoric animals, called dire wolves, were large and must have been fierce hunters.

like other dogs (canines), has blunt claws. Its feet are suited to running, not to climbing.

Dire wolves date back to Pleistocene times—an age of cave men and such monsters as saber-toothed tigers and mammoths. The dire wolves must have been fierce competitors for the early hunters. Their fossil remains show them to have been large and powerful. Probably the prehistoric people developed a hatred for them which never died out and ancient dislikes as well as myths were carried along to become associated, eventually, with *Canis lupus,* the beautiful timber wolf of today.

Although horror stories about wolves are still popular in Europe, the animals long ago ceased to be a threat there. Conditions were different, however, in medieval times. Human populations were steadily increasing then, and people were beginning to make their living on farms. Wolves were numerous and their food requirements were enormous. As a result, there were constant clashes between men and beasts.

Gradually humans got the upper hand. By the early part of the sixteenth century, wolves had been completely exterminated in England. In another hundred years they were gone even from their mountain retreats in Scotland. By the eighteenth century none were surviving in Ireland.

On the continent the persecution increased along with the use of firearms. In France the militia frequently was employed to drive the animals from wooded areas so they could be shot easily by gunmen waiting in open areas.

Nevertheless, some small wolf packs managed to outwit their enemies. Today they are known to live in mountainous parts of France, Spain, Italy, and on the Balkan peninsula. They are even more common in parts of northern Asia. "Wolf attacks" continue to make news from time to time. However, when such reports are published, and someone takes the trouble to investigate, it turns out that the attacking animals were wild dogs rather than wolves.

Although this picture was taken in a modern zoo, the atmosphere suggests those long-ago days when wolves were numerous and roamed in the forests of Europe, Asia, and North America.
Wide World Photos

The war against wolves was carried to North America by the earliest settlers from Europe, who came to the New World bringing all the fears and distrust which had existed in their native countries. As they began settling wilderness areas, they found new wolf problems. "Ye devouring wulff," it was reported in early Colonial times, constantly raided farmlands, destroying sheep, pigs, and cows—doing more damage than all other predators combined.

The situation was almost beyond control because the crude guns owned by the settlers were not satisfactory for wolf-hunting. And other means of thwarting the raiding wolves, such as wire fencing, were not known then.

One device commonly used was the wolf pit. This was a large, deep, circular trench, with a central area left at ground level. Bait would be placed on this, and the trench would be concealed by vegetation. The wolves, rushing toward the bait, would fall through the branches and leaves, into the pit. As a victim was trapped in this manner, a farmer would not waste valuable ammunition by shooting. He simply climbed into the pit and bludgeoned the animal with a club or axe.

Sometimes a man undertook this bloodthirsty job when several wolves were entrapped. It seems unlikely, therefore, that the animals were feared for their attacks on humans.

For years an organized war against wolves was carried on, sparked by the payment of bounty. The Pilgrims began it only a few years after reaching

New England. Soon after, there was a bounty system in Virginia. Money was paid to Indian trappers and to some eager hunters who made a career of destroying wolves. They did not concentrate on adults, but sought out dens and killed entire families of cubs.

In time farmers began to use poison to eliminate these animals that did not sense the difference between their natural prey and the livestock that had been introduced into their country.

But in spite of all the campaigns against them, wolves continued for a long time to thrive in North America. As pioneers pushed westward, they were found everywhere on the continent. Ultimately, it was a combination of pressures that turned them into losers. With the clearing of countless stretches of forest, and the employment against them of guns, poisons, and traps, they were overwhelmed. Today timber wolves are found only in small areas of Michigan and Minnesota. A smaller species is on the verge of extinction in Texas and Louisiana.

Canis lupus has a variety of popular names. Besides timber wolf, it is known as gray wolf, buffalo wolf, and barren grounds wolf. Despite the "gray" name, the species has a variety of color shades, from white to almost black. Some are brindled brown or yellowish. In their northern Canadian range, individuals may weigh a hundred and seventy pounds or more. In more southern ranges, the average weight is lighter, often under a hundred pounds.

Canada has a somewhat healthier wolf popula-

Timber wolves (two young ones are shown)
are greatly admired by people who study
them. There is not a single proven instance
of a North American wolf attacking a human.

tion than the United States. But there, too, the
numbers have been dwindling rapidly, so that government agencies have stopped encouraging their
destruction. Wolf-hunting, however, is carried on as
a "sport," the shooting often being done from small
aircraft after the animals have been flushed into the
open by giant firecrackers.

Disputes go on endlessly between conservation-
ists and the ranchers and farmers as to whether

wolves should be completely exterminated or given the protection that is essential if they are to survive. Biologists make these points:

Wolves actually should be considered helpful in keeping game herds at their best population levels. They prey chiefly on sick individuals and on old animals that are past the breeding age. Human hunters, on the contrary, are apt to aim at those in the prime of life.

Any danger to humans from wolves is most unlikely.

If an individual wolf leaves its usual habitat and starts preying on livestock, it can easily be eliminated. . . . For their own good, wolves should not be allowed to live close enough to ranches or farms so they can easily carry out raids, because at the first sign of trouble, there is a demand for their extinction.

What is the true nature of a wolf? The animal has been studied closely by many scientists. All of them gave admiring and even affectionate descriptions of this creature which fiction writers like to call "monster" and "bloody beast."

In their own society wolves lead a model family life. Usually when a male and female choose each other as mates, they remain together until separated by death. Although they spend much of their time in the open, a pair always finds, or digs, a den which will shelter their expected family. After the young have been born, and have been weaned from the mother's milk, the father is very concerned about bringing them food. He also watches out for their safety and, in time, helps the mother train them to fend for themselves. This close family unit remains together for two, or sometimes three, years.

All adults seem to have concern and affection for the cubs. It has been observed that if something has happened to detain parents from returning to their den with food, some neighbor wolf will step in to take on the job of "baby-sitting."

The term pack is used so often, there is a general

A wolf pet is a challenge, even though it behaves very much like a domestic dog. It needs more space for exercising, more food, and, because of its great strength, more attention from adults.

Wide World Photos

impression that, by choice, wolves travel in large groups. However, to scientists, as few as three individuals occupying a definite territory and hunting together may be considered a pack. The average pack, when the animals are free from outside interference, is made up of about six. Quite different from the great, ferocious crowd of a hundred or more that are described in many a fiction story!

A "lone" wolf is an exception. Probably such an individual is old—too old to run with a pack, and its teeth are in poor condition. Therefore it follows others at a distance and eats any game they might leave behind.

A wolf's need for companionship is remarkable. If deprived of association with other wolves, it will accept human friendship with pleasure. Taken as a young cub, one will make an excellent pet—under certain conditions. The owner should understand that a wolf needs enormous amounts of food and, fully grown, is so big and strong it must live in a suitable environment with adults.

Even wild wolves show a friendly interest in people, when not being hunted. A number of stories have been told by reliable reporters to prove the truth of this statement. One concerned Dr. George Goodwin, a mammalogist on the staff of The American Museum of Natural History.

Dr. Goodwin was doing scientific research in Canada's far north. Alone one night, he decided to not bother with any shelter but his sleeping bag. He had a fresh kill near the spot he settled on, and after darkness fell, wolves closed in—no more than ten feet from him. There they stood still, looking at him with interest. In a short time they ran away, without having made an unfriendly move or trying to steal the kill that had not been of their own making.

A number of zoologists have investigated reports of people being attacked by wolves. All came to the same conclusion: In North America there is *not one* proven case of a human being killed, or even attacked, by a wolf.

In Europe and Asia, long ago, there probably were wolf attacks, possibly because the animals were starving and men on horseback could provide desperately needed food. Another factor was rabies. The dread disease was common among animals in Europe, and a rabid wolf would not be normal in its behavior. It could indeed be the "mad attacker" portrayed in horror stories.

Besides the reputation inherited from their unfortunate relatives in Europe, North American wolves in general suffered from the actions of certain notorious individuals such as the "Custer Wolf of the Black Hills," the "Phantom," and "Dakota Three Toes." As with human troublemakers like Jesse James, these characters were described on posters and rewards were offered for their capture—dead or alive. They were wolves that, with extraordinary intelligence and cunning, were able to outwit all their enemies. Dakota Three Toes, for example, was WANTED for fourteen years. During all that time, as he was being hunted, he destroyed thousands of dollars worth of livestock.

Even without worry over livestock or personal attack, anyone who hears the wolf's evening call—a deep, long-drawn-out, tremulous howl—is likely to feel a shiver of fright. Yet the call is not aggressive; it is conversation among friendly wolves. The mating call is a throaty howl. The call of the chase is a deep, guttural roar. Some calls are given during daylight hours, but more often they are sounds of the night.

Among the most famed of animal sounds is the howl of the coyote, often called prairie wolf. As pioneers moved westward in North America, they had to calm their nerves against the wild coyote howls that soared over the prairies. They seemed eerie and sinister, but the people learned that one animal could sound like a whole pack, and that the coyotes were not challenging their human neighbors.

An evening howl is a ritual. It has a haunting, unforgettable quality, and carries for a remarkable distance on the still night air. Like their "first cousin" wolves, coyotes have several distinct calls. Apparently one type warns the pack of danger and others ask for assistance or information.

The more we learn about wolves, the more we can admire and respect them. Nevertheless, ancient prejudice and superstition linger on, and the "big, bad wolf" continues to be a symbol of menace and destruction.

2
The Great Cats

On one page of a daily newspaper recently were two headlines: LION MAULS MAN IN JUNGLE HABITAT and DAD RESCUES DAUGHTER FROM JAWS OF A TIGER.

This paper and others carrying the same stories would reach millions of readers, and many of them would think, "Those vicious animals! Always trying to attack people!"

But anyone who read the complete stories would see it another way, with the animals not deserving all the blame. In the first case, a man had been driving through a new type of zoo—an area where African mammals are given complete freedom. People were not allowed to walk there, and were told to have their car windows closed at all times. This visitor had a window open, even when the car was slowed down by traffic. A lion, obviously investigating the strange intruders on wheels, slapped his paw through the opening, and so mauled his victim.

The tiger in the story was being kept in a pit at a zoo. A visitor accidentally dropped a picture into the pit, then tried to pick it up with a long, pointed stick. The young, agile tiger became so upset by having this poking around him, he leaped up and cleared the top of the pit. He happened to land by a mother and her daughter. The mother was too frightened to move, but the little girl started to run; whereupon the tiger grabbed her in his mouth, and ran even faster. That was when her father appeared. He caught the tiger, clamped a wrestling hold on him, forcing him to free the child.

Vicious? These were bewildered or frightened animals. They were not attacking people simply because of a natural hostility.

But then we hear of the man-eating tigers of Indian and other jungle areas. Are not the stories of terror caused by such beasts true?

They are indeed. Records kept in India show that, over a long period of time, human victims of tigers amounted to an average of from one thousand to two thousand victims each year.

The reasons for tigers becoming man-eaters vary, as the animals themselves do. An expert has said, "There is no such thing as *the* tiger; every tiger is an individual." And along with stories of their ferocity are many accounts of friendly tigers. Captive cubs become devoted to their keepers and many remain affectionate pets as adults. There are even instances of a captured adult allowing his keeper to stroke his fur only a few days after the capture.

A tiger's weapons, as with all carnivores, are four canine teeth—two in the lower jaw fit in front of two in the upper jaw. Long and sharply pointed, these can puncture and slash as if they were giant knives.

In having this equipment, tigers are like wolves, but their hunting methods are very different. Wolves chase their prey. Tigers stalk, then spring. They are extremely agile, and can cover as much as fifteen feet in a single leap.

Tiger cubs are as playful as pet kittens, but this "play" is likely to be rough. The cubs are too strong to be gentle in their touch.

Wide World Photos

The great cats have the largest and sharpest canine teeth of all flesh-eating mammals. The accidental breaking of one or more canines can mean starvation for a tiger or other member of the feline family. .

Tigers do not have the bone-crushing teeth of wolves, but the upper surface of the tongue is so rough it can draw blood merely by licking the surface of human skin. They also have the sharpest claws of any mammal. Normally these are kept in sheaths in the paws, but they can quickly extend for action.

The size of tigers varies, from the huge, long-haired species of Manchuria (thirteen or more feet in length, including tail) to the average Indian specimen of about nine feet, to the smallest, which lives on the island of Bali. A really large tiger may weigh more than six hundred and fifty pounds.

Naturally animals of this size require large quantities of food. Even in a zoo, where the big cats' only exercise is a limited pacing back and forth, an average tiger needs about ten pounds of meat a day. In their native jungles they must have plenty of wildlife or they will seek out cattle and other domestic animals. When this happens, conflict between people and tigers is likely to begin.

But this is not always the case. In many areas of Indochina and China the great cats help to protect crops from wild pigs, deer, and monkeys by preying on them. If they occasionally attack domestic cattle, the farmer does not object; this is well deserved payment for "services rendered." Such tigers rarely bother humans, and sometimes become family pets.

We must see these "friendly" tigers in a different light from those persecuted by hunters. A tiger that has been shot at quickly becomes wary—and smart. Afterward, whether or not he was wounded, he is likely to attack anyone he finds trespassing on his territory.

It seems rather strange that such great flesh-eaters do not normally prey on humans. An explanation may be that long before men began to invade the jungles, tigers were set in their food-hunting habits, and their prey consisted of four-footed animals that scampered or crept through the underbrush. The upright stance and unique movements of a human would arouse a tiger's curiosity more than suggest an easy meal.

However, during the eighteenth and nineteenth centuries, the world of the tiger was changing. Until then, the natural habitats had been undisturbed by civilization, and animal life of all kinds flourished. Then there came a great increase in the human population, and the needs of people were a threat to the welfare of the animals.

Workmen tapping trees for rubber and gathering lumber or firewood were targets for the big cats when game happened to be scarce. Soon the killing of a man by a tiger was not a rare happening. Reports show that in 1769 more than four hundred people lost their lives to tigers in one Indian village. During the years that followed, as roads were built and telegraph wires were put through jungle areas, many more such horrifying statistics were recorded.

Man-killing tigers are not necessarily man-eaters. Many times one of them has been known to attack a person when frightened, or because he was suffering—perhaps from a bullet wound, a thorn in a paw, or festering sores caused by embedded porcupine quills; but the body of the victim was left untouched after the killing. A young tiger, supercharged with energy, may leap at someone just because that individual happens to be close at hand.

An important cause of man-eating is the disability of older tigers. As a tiger ages and its teeth wear down, it is less able to deal with natural prey; humans are easier victims. And a mother with cubs to be fed may be desperate for food, going after any form of animal life—human included.

Natural disasters such as drought, plague, or famine have sometimes been followed by a serious increase of man-eaters. Not only would their usual jungle prey become scarce, but many human bodies, left unburied, would be theirs for the taking.

After becoming accustomed to human flesh, tigers are likely to be a permanent threat to the humans in the surrounding area. Appalling stories have been told of man-eaters that became bold enough to enter villages and break into huts to carry victims away. In some districts of India, so many natives have been killed in this way that the survivors finally deserted their homes and searched for a new locality where, hopefully, the big cats would not find them.

Although tigers of this sort have an evil reputation, the leopard is a still more savage cat. It can fight as well as tigers and is more given to climbing. As a result, it may leap from a tree on unsuspecting victims. Also, it is intelligent—quick to learn and apparently able to remember anything helpful to its well-being. Normally it does not attack people. But in leopard country, dog owners must beware; leopards have a fondness for the flesh of domestic dogs.

By contrast, the lion is a gentle animal. As people, years ago, were becoming acquainted with jungle animals, this so-called "king of the beasts" was held in awe throughout the world. His tremendous roar, his massive head and handsome face, often set off by a luxurious mane, his huge body—these features added up to a regal, though frightening, figure.

Then lions began to be captured and brought from their natural surroundings to be exhibited in zoos and circuses. Under such conditions they often seemed treacherous, snarling characters, eager to attack their keepers or trainers. Show people never discouraged this thinking; the more fierce an animal seemed, the more exciting an act would be.

Today lions have a new image, much of it due to the story of Elsa, the lioness who was raised in East Africa by Joy and George Adamson. *Born Free,* and other books about Elsa and her cubs, gave millions of people a new understanding of lions.

They are rather friendly beasts. When well fed they are relaxed and lazy, and not inclined to attack other animals just for the sake of killing. Unlike the solitary tiger, a lion travels with a group, or pride. When actually hunting, one may work alone, or two or more may team up. However, once prey has been secured, those involved in the hunt do not fight over it, but feed contentedly, enjoying what appears to be a family dinner.

Although the Adamsons brought Elsa up as a pet, they taught her to kill so that she could fend for

herself when fully grown. Nevertheless, her affection for her human "parents" was so strong she kept in touch with them long after she was living free and had mated with a lion who had never been associated with people.

What of man-eating lions?

Authentic stories of them have been recounted, so we know such killings do occur. Usually, as with tigers, any attack against a human is made by an old individual that finds it hard to catch more active prey.

A fearful example of man-eaters was the ravaging of railroad workers when the Uganda Railway was being constructed in Africa at the end of the nineteenth century. Two male lions began a reign of terror by killing sheep and goats that were being kept by the workmen. Then they turned to people. Night after night one or the other would steal into camp and seize a victim. It would be done too quietly and quickly for others to know until a great roar came from the nearby forest. "Satan has struck again," the native workers would lament.

Within nine months scores of people had met death in this way. All the resources of the British Empire were unable to stop those two lions. Finally work on the railroad was halted while every possible

Wide World Photos

Lions are not as savage as some other big cats.
Unless hungry, they are lazy rather than fierce.
A large male, with a full mane, is a regal figure.

device was used. After a campaign of several weeks, the lions were tracked down and shot. The railroad project was then continued.

Man-eating lions still appear occasionally, but with modern weapons, any such threatening animal is quickly killed. Today, however, people are thinking more about saving lions than destroying them, and on safaris they enjoy photographing, rather than shooting, the great cats.

Long ago lions were common in Asia and parts

Lion cubs are nursed until three months of age. During this time the father often brings food to his mate and may continue to supply the family with game for an even longer time. Cubs are not able to kill for themselves until they are about a year old.

Mark Boulton from National Audubon Society

of the Middle East, as well as in Africa. But now, outside of Africa, fewer than two hundred remain, and these exist only because they are given government protection in the Gir Forest of India.

Tigers, like lions, have almost reached the vanishing point. Once abundant from the tropics of India north to Siberia, in the 1970s fewer than two thousand wild tigers remain in the world. Conservation laws offer them protection from hunters and fur dealers, but poachers, who can expect to be paid more than a thousand dollars for a beautiful "illegal" tiger skin, carry on the grim work of extermination.

3

Bear Hugs– Fatal or Friendly?

A "bear hug" may sound like an affectionate kind of greeting. But it isn't really. According to an old belief it would actually be a grip of death. Bears were said to kill an opponent by hugging or squeezing the life out of him.

"Not so!" say scientists who know bears. A bear's method of attack is to strike with its paws, driving the claws into the victim's body.

Bears can be deadly, though not because of their hugging.

Many people feel there is nothing frightening about a bear. In contrast to the prejudice built up against wolves by age-old folk stories, bears are usually presented as delightful, amusing creatures. The cubs have tremendous appeal, and even fully grown bears look as if they would be fun to play with.

However, if we encounter one of these shaggy

characters in a wilderness area, we may form a different opinion. Suspicious of a stranger or disappointed when "handouts" of food come to an end, a bear may become ferocious. Then a hunter or an innocent camper becomes the victim of a sudden attack.

The grizzly and the American black bear are the species campers must watch for in national parks. They are curious and ever-hungry. As a result, food supplies—unless carefully protected—are likely to disappear during the night. Not only are packages demolished, cans are ripped open with more speed than an electric can opener could manage.

If bears find "good pickings" when campers are asleep or away from their supplies, they may become bold enough to come back, despite daylight and the presence of people. This is when they become really frightening.

In prehistoric times bears must have terrified the primitive people of Europe. They were gigantic—larger than the immense Kodiak bear of today. These animals surely must have competed with humans for food, even if they did not plan to add people to their diet. Their fossilized bones have been found in the same remains with crude spears and hatchets.

Bears have a wide range of eating taste. They are classed as carnivores because they devour fish and game. But they also feed on roots, grass, nuts, and fruit, and have a particular fondness for honey.

The "game" they seek is likely to be small, something like mice and ground squirrels. But any of the big bears can easily crush such mammals as sheep or elk. A grizzly is apt to do this if any of them stray into his territory.

Some years ago grizzly bears were recognized as masters of American northwest wildlife. They were much feared by other animals and by explorers, fur trappers, and pioneering cattlemen. Though they are not the biggest of bears, an average individual weighs about five hundred pounds.

A dreadful "sport" that was popular among Spaniard colonists in Old California was bullfighting, using a grizzly bear instead of a human against the bull. Usually the bear was victorious. It was not uncommon for a single bear to kill as many as six bulls in one afternoon.

Grizzlies may be just as deadly to anyone they encounter in their native woodlands. It is agreed that the best way for an unarmed person to survive a meeting is to "play dead."

A number of fanciful tales have been told about grizzlies over the years. One concerns a habit of biting and scratching trees. This, it was claimed, was a challenge to would-be rivals: The bear bit a chunk of bark from a tree trunk and reached up to scratch above it to show, in effect, how tall he was, and thus establish himself as the boss bear of the area until a bigger bear came along.

Responsible naturalists have reported seeing a grizzly stand up with his paws against a trunk, sniff,

These grizzly bear cubs, only three and a
half months old, already have claws that
can mean trouble for either man or beast.

then tear off a chunk of wood with his teeth. However, the act seemed to be carried out in a spirit of
wanting something to do; there was nothing challenging about it.

As with other kinds of animals, grizzlies have no
defense against high-powered rifles or against the
loss of their natural habitat. Today only a few remain in the United States, and they are within the
confines of the national parks. Survival has been better to the far north, and fair numbers still exist from

British Columbia to Alaska. They were wiped out long ago in Europe.

Black bears are smaller than the grizzlies. A male has a body length of no more than six and a half feet, with a weight between two and three hundred pounds, although an occasional individual is larger and heavier. They are great tree climbers, and they use this ability to go after bee nests in elevated branches, since they are very fond of honey! They also are strong swimmers, and will cross swift-flowing rivers or lakes several miles wide.

The common "black" bear may not always be recognized by its color, for there are several variations. The fur of the cinnamon bear, for example, is brown, like the spice. But in all other ways it is like the black bear. Sometimes a "cinnamon" mother produces both black and brown cubs.

Although black bears have the ability to kill a human as easily as a grizzily can, they are less aggressive. In fact, when not frightened, they are inclined to be friendly and, because of their curiosity, may be easily captured when young. But the pet black bear is not a safe companion indefinitely. After its second birthday, it may suddenly become irritable and unpredictable and attack the person who has been giving it affectionate care.

Differences that may be noted in grizzlies and black bears are: The grizzly has a pronounced shoulder hump that is missing in black bears. Its head is broader and larger in proportion to its body than the black bear's. It is held lower, as the bear trudges along. The claws of a black bear are shorter, more curved, and especially sharp, making tree-climbing possible despite the lumbering body.

A black bear of Asia is smaller than the American species and, because of a white crescent on its chest, it is known as the moon bear. It is not popular with people of Burma, China, and other countries to which it is native because it raids orchards and destroys beehives. It will aggressively attack humans, and has even been accused of eating children. In

Black bears in a national park appear very friendly as they come upon people and beg for food. However, visitors are told not to feed them because they may suddenly become irritable—and very dangerous.

Grant Haist from National Audubon Society

China it has a variety of local names, including "man" bear, "dog" bear, and "pig" bear.

Also living in Asia is the smallest of bears—the sun bear, which usually weighs no more than a hundred pounds. Its name refers to a splash of orange or white on the chest which supposedly represents the rising sun.

Sun bears live in dense jungles. They are expert climbers and most of their hunting is done in tree branches, where they look for fruit and birds' nests. They are easily tamed and, because of their convenient size and playful disposition, they are popular as pets. But as with other bears, their behavior may be unpredictable. In a natural state, an adult will fight ferociously when cornered.

Anyone inclined to be frightened by size alone would be most terrified by the brown bears of Alaska, British Columbia, and the islands in between. They are the largest flesh-eating animals in the world. The Kodiak, which is one variety, may have a height of nine feet. When standing upright, it may be twelve feet tall!

These are exceptional measurements. But some Alaskan brown bears confined in zoos become very fat, and may get close to the two-thousand pound mark on the scales.

Courtesy of the American Museum of Natural History

Great Alaskan brown bears, standing upright, may be as much as twelve feet tall. They eat vegetation, insects, and small mammals. But when spring comes, they then feast on salmon.

Interestingly enough, these great animals, although classed as carnivores, do not eat as much flesh as do smaller mammals, such as wolves. They feed mainly on vegetation, grazing like cattle during the spring. Later in the year, insects and rodents are added to their diet. June is their feasting time, when salmon come up the northern rivers to spawn. After spawning, the fish die, and the bears are ready and waiting to devour their remains. Scientists have suggested that this fare, rich in minerals and vitamins, is responsible for the great size attained by bears of the region.

The largest bear of the Old World lives in Siberia and it too feeds on salmon after the spawning season.

There are other brown bears of Asia and Europe that are better known because they are frequently seen in zoos and circuses. They become excellent performers—dancing, wrestling, riding bicycles, and doing other stunts—and usually seem to enjoy having an audience. However, trainers learn to be wary. A bear may become angry without giving any warning.

Strange myths have existed through the years about almost every kind of animal, with none more fantastic than some beliefs about bears. One of these claimed that cubs were born as shapeless bits of pulp, and that the mother at once got busy "licking them into shape." This meant forming them with her tongue as an artist would shape clay with his hands. With constant licking, the bits of "pulp" would develop legs and head and, in general, shape up as a recognizable bear cub.

The inspiration for such an incredible story can be understood: Hunters and casual observers of wildlife glimpsed young cubs—and saw furless little blobs that bore no resemblance to the handsome form of a grown bear. They also noted that mother bears seemed to lick their offspring endlessly. Putting these two pieces of information together, they arrived at a completely wrong conclusion. But it was believed by many people; it was even printed in some of the earliest books on natural history, written nearly two thousand years ago.

Today the life story of bears is no mystery: Cubs are born during the mother's long winter retreat in her den. They are tiny. A two-hundred-pound female may give birth to young that weigh less than a pound. In the spring the cubs leave the den, and soon follow their mothers in a quest for food. Male bears are not involved with their families; they go off on their own after mating.

But even though basic facts about the life of a bear are well known, the animal is still involved in a nature mystery—that of the Abominable Snowman.

The legend of a great, hairy, ape-like creature began years ago and has not been allowed to die. Periodically, people claim to have seen signs of one— usually giant footprints.

Investigators note that when a bear lumbers

along, walking only on his hind feet, the soles are pressed on the ground, and there are five toes, as with human feet. Footprints left in the snow in remote areas, such as the slopes of the Himalaya Mountains, may enlarge as some melting occurs, then freeze again. In this way a bear's tracks may come to resemble the footprints of a superman, and so cause speculation about a monster never seen "in the flesh."

4

Gorilla –
The Perfect Monster

A "monster" may be described in various ways, but any description is likely to include such words as hideous and huge, powerful and frightening.

Such adjectives seem well suited to a gorilla. This animal is heavily built, with enormous arms that reach almost to the ground when it is standing upright. A large male, five and a half feet tall, may weigh five hundred or more pounds, and have an arm-spread of eight feet. And the face! Massive jaws, sunken eyes, large nostrils, and huge teeth combine to form the image of a real ogre.

Some of the actions of gorillas and the sounds made by them add to the impression that they are ferocious beasts. They beat their chests in what appears to be rage and defiance, doing this even in captivity. And a male may produce a mighty roar—a series of snarling barks, followed by a deep rumble.

Besides these facts, fiction writers have portrayed the great apes as vicious, violent, and sinister personalities. The name King Kong, a fictitious gorilla, has come to mean brute force and cause for terror.

Fantastic ideas about gorillas date back many years, to the time people were first becoming interested in jungle animals. The earliest reports about them were given by travelers and missionaries to Africa. Undoubtedly, firsthand descriptions were few; reports were built on stories told by natives.

A description given by a missionary little more than a hundred years ago spoke of gorillas being "exceedingly ferocious" and "objects of terror." Another wrote that the gorilla was "one of the most frightful animals in the world."

This was quite a reputation to live down! And it was many years before anyone made a serious effort to find out the truth about these apes. Doing so was not easy. Even at the end of the nineteenth century, when a zoologist went to Africa especially to study them, his methods proved unsatisfactory, although he had a better approach than previous explorers. He did not crash through the underbrush with a gun ready for instant action. Instead, he built a large iron cage in the jungle and installed himself in it to quietly watch the animal life around him.

However, the several months he spent there did little to increase an understanding of gorillas. The great apes, gorillas and chimpanzees alike, stayed away from the man and his cage. He heard screams, roars, and thumping nearby, but had no way of knowing what animals were responsible.

In the years that followed, people's contacts with gorillas were chiefly through men who shot them for "sport" or for museum displays of stuffed animals, or who captured them for exhibition in zoos or circuses. Those that were taken alive did little to help in understanding others like them. They were kept in surroundings completely unnatural to them, and were often mistreated because of ignorance and fear on the part of their keepers. And since an audience was usually thrilled to see a ferocious wild beast rather than a peaceable one, circus managers presented their captive gorillas, so far as possible, in the King Kong image.

It was not until 1959 that someone came along who dispelled misunderstandings and superstitions about the great beast.

George Schaller, a zoologist, decided the only way to observe them properly was to live in their territory, adopting their way of life. Consequently, he and his wife, for two years, made their home in a small wooden hut in a gorilla-occupied African jungle. Day after day, without a gun or other weapon, he followed groups of gorillas, just watching. Often he would not return to his hut at night but slept outdoors with branches for a bed, just as the apes were doing.

The gorillas accepted him! When they first became aware that "something" different was in their

The "ferocious" gorilla of fiction proves to be
a peace-loving animal. Here a mother munches
leaves—a favorite occupation of gorillas—
while her baby thinks about trying some.

midst, they were cautious, even suspicious. But they
made no unfriendly moves, and soon realized he was
harmless.

Mr. Schaller tells of one encounter when he was
sitting on a branch about five feet above the ground
observing a group of twenty-one gorillas munch on
leaves. Then they would rest, and munch some
more. After a couple of hours, they noticed him, and
all walked over for a closer look. They seemed ready

to turn away, but suddenly one pulled itself up on his branch and sat down next to him. Man and beast stole glances at each other. What was this strange "creature" really like?

Finally the gorilla's curiosity was satisfied. With a long, powerful arm, it swung itself back to the ground and moved away. This was not the end, however. Another member of the troop and then another and another took turns sitting on the branch with the astonished scientist.

Since Mr. Schaller's studies, others have been carried on in the gorillas' habitat. Dian Fossey, a student of animal behavior, recently completed a five-year stay in the mountainous regions of central Africa, and there became completely accepted as a friend and neighbor by more than a hundred of the apes. She would sit with a troop and pretend to eat leaves with them. Young ones played around her with no objections from their mothers.

Altogether, gorillas showed themselves to be peace-loving rather than aggressive. The chest-beating, the hoots, and the roars are not related to a bloodthirsty disposition.

One important function of chest-beating is to help members of a troop locate each other. The booming *pok pok pok* sound that a full-grown male produces carries a long distance, and if a troop has scattered during feeding time, it determines where the leader is.

Chest-beating may also keep a rival gorilla or

possible enemy at a safe distance. It also serves no-
tice that a leader is on the alert and wants no interference.

The social groups in which gorillas live usually
are made up of females and young individuals, including some males which wander in and out of the
friendly circle. And every group, or troop, has a
leader in the form of a big silverback.

Silverbacks were given this name because when
a male is about ten years old, silver-gray hairs begin
to appear on his back. They grow more numerous
until a gray saddle is formed in the midst of his
shiny black hair.

Roaring and chest-beating are the specialties of
every leader. When he feels threatened, they become
part of a dramatic routine. In the first act he inflates
air sacs in his chest. Then, from either a sitting or
standing position, he begins to hoot.

Suddenly he stops the noise and places a leaf between his lips. After holding it a moment, he opens
his mouth so that it falls out. Then he begins hooting again, faster and louder than before. Again the
hooting stops, while he pulls himself to his full
height and slaps his big hands against his chest and
belly—and the sound is like a booming drum.

Now he is ready for fast action. Upright, he runs
sideways, then drops on all fours and dashes for-

Courtesy of the American Museum of Natural History

A male gorilla exhibits a great show of force by
beating his chest. But this is mostly a dramatic
bluff. The mighty ape is really not aggressive.

ward, snatching at vegetation and tearing branches off trees and bushes and tossing them into the air. As he runs he slaps at anything in his way, including other gorillas. The outburst of violence ends as he thumps the ground with the palms of his hands.

If this display is directed against another species of animal or a human, the final act seems to depend on the action of the opponent. If the "intruder" refuses to be frightened by the fierce display, the gorilla is likely to calmly lead his troop from the danger zone. On the other hand, if the intruder starts to run, the silverback may run after him on all fours, and, catching him, bite and scratch.

Many African natives have been wounded in this way while gorilla-hunting. But they are not killed. The powerful ape leaves his mark on the enemy and lets it go at that; he does not pummel or maul him.

Of course, if his adversary is armed with a loaded gun, the gorilla stands no chance of settling things his way. Mighty as he is, a single bullet can send him crashing to the earth.

Adult females also beat their chests, but they do not hoot or carry out the dramatic warning routine of the silverbacks.

Chest-beating apparently is instinctive to the gorilla nature. Young ones growing up in captivity, that have never seen adults perform, sometimes show defiance by such actions.

People are apt to wonder if gorillas fight among

themselves. Mr. Schaller feels this happens very seldom. Members of a troop are well-behaved as a rule. Two angry females may start to scream at each other, but if they do, the leader glares at them from his deep-set eyes and they quickly grow quiet. And if young ones get too rough in their play, any adult may bring order by slapping the ground with a determined thump.

From his observations Mr. Schaller decided that gorillas of different troops do not enjoy fighting about territory, or for any other reason. He feels they actually try to avoid confronting each other.

It is not only behavior in their jungle homes that shows gorillas to be more peace-loving than aggressive and more affectionate than brutal. Individuals captured when young have lived for years in some of the great zoos of the world, and responded warmly to attention from their keepers.

There have been a number of famous captives. "John Daniel," a young male, lived in the home of Alyse Cunningham in London from 1918 to 1921. When Mrs. Cunningham was forced to give him up, and he was brought to the United States, he refused to eat, and died within a matter of weeks. The diagnosis: "Of a broken heart."

A young female named Toto became a pet of Maria Hoyt, and the Hoyt family did everything possible to give the little orphan from the jungle a good home. Toto responded enthusiastically. She even made up for not having babies of her own by "adopting" other pets in the family, such as a young cat which she carried around like a happy mother.

But as with any gorilla (or any large "wild") pet, there were problems, due to her tremendous strength. A playful push from a muscular animal weighing hundreds of pounds can be a disaster. After nine years, Mrs. Hoyt felt that handling Toto was too difficult, and sold her to the Ringling Brothers' great circus. But a few years later she bought the gorilla back again, and was devoted to her for the next twelve years, until Toto's death.

However, in spite of these and other well-known stories about the gentle giants, the hair-raising myths continue. And it cannot be denied that, going by looks alone, a gorilla does make a "perfect monster."

5

Dracula
Was Not a Bat

Animals do not need to be big to be alarming. To countless people one of the most frightening of all is quite small—the bat.

Some fears are understandable. Bats are expert fliers, and if they invade a barn or house, their swooping bodies are very unsettling to human occupants.

Bats are equipped with a natural "radar" system, which makes it unlikely they will ever bump into anything. Nevertheless, there is a widespread belief to the contrary. Many people think that if one of these flying mammals is in an enclosed area with a human, it will surely become entangled in the person's hair.

Even more disturbing is the knowledge that some bats need blood for nourishment. Here is the basis for endless horror stories! Bloodsucking bats have been involved in so many fantasies, it becomes difficult to remember actual facts about them.

The most gruesome of tales are concerned with vampire bats which, in turn, are concerned with mythical "corpses" that rise from their graves at night to go in search of human blood. Such tales first originated in Europe hundreds of years ago.

In 1897 some of these ancient superstitions became the basis for a book called *Dracula*. In it the author, Bram Stoker, created the character of Count Dracula, a "vampire" who lived on blood. Dracula was a man of commanding appearance, deathly pale except for very red lips; and he had unusually sharp, pointed canine teeth. By night he assumed the form of an animal, such as wolf, rat, or bat, and he restored his vitality by biting humans and sucking their blood—at the same time turning his victims into vampires!

Count Dracula was referred to as "half human, half bat," and the tremendous success of Mr. Stoker's story added many hair-raising, mystical ideas about bats to the folklore that already existed. But of course the idea of human vampires was completely imaginative, and we cannot blame the misdeeds of the old Count on our flying mammals.

Strangely, true vampire bats are not found in Europe where these legends originated. They live only in tropical or near-tropical climates of Central America and southward to Brazil. When blood-nourished bats were found in South America (Charles Darwin discovered them during his historic explorative voyage on the *Beagle* some years before

Dracula was written), vampire seemed the most suitable name for them because of the Old World superstitions.

A vampire bat in appearance is nothing like as fearful as some other species of this mammal. It is small—barely three inches long, with a wingspread of about fifteen inches. And it does not have the "nose leaf" that gives some bats a grotesque and hideous look.

The teeth give a clue to the animal's unique diet of blood. There are two incisors at the center of the top jaw. They are large, curved, and extremely sharp. The canine teeth next to them are long,

This enlargement of a vampire bat's head
shows the sharp teeth that are its cutting tools.

Charles Mohr from National Audubon Society

pointed, and sharp. All other teeth are small and unimportant.

The two incisors and two canine teeth do all the gory work needed to sustain life. With the skill of an expert surgeon, they slice the victim's skin. The victim, if sleeping at the same time, may not be disturbed enough to waken. Once the incision has been made, the bat does not actually suck blood, but rather sips it. The tongue is stuck out and pulled in with short, quick strokes, and blood is drawn into the bat's mouth beneath it.

A human is most likely to be cut on the lips or ear, forehead or finger. Other mammals are more apt to be attacked on the neck, nose, or ear. Such victims do not usually suffer a dangerous loss of blood because only a small amount is taken. If the wound heals cleanly, no harm is done. But a small bird may lose enough in proportion to its size to die in one night's "bloodletting."

Vampire bats can inflict something more serious than a small loss of blood on any victim: Rabies is frequently transmitted by their bite. Unless it is treated promptly, this frightful disease usually causes death. However, vampire bats are able to contract it and still live. Therefore, many rabid individuals exist, and as they go after the blood they continually need, they spread rabies to their victims.

Because of the possibility of rabies, a fear of all bats is widespread, and it is true that even bats that live on fruits or insects may spread the disease. However, a bite from a member of such species is very rare, even in areas where the flying mammals live close to human dwellings.

For a number of years, in a large area extending from central Mexico to central Argentina, vampire bats have been a serious problem to ranchers. With numerous horses and cattle introduced to this region, the bats had a great supply of blood easily available, and their population increased to an alarming extent. As they were preyed upon, the livestock began to weaken noticeably. Beef cattle lost weight and a cow that normally yielded thirty quarts of milk a day produced no more than twenty. Horses became sickly.

Very recently, after intensive studies and work, scientists have discovered a way to defeat the vampires. Their "secret weapon" is an anticoagulant—a chemical that prevents blood from clotting. The scientists trap a few bats, daub them with the anticoagulant, and release them. The little mammals then fly back to their roosts and begin to groom their fur by licking it. Other bats of the colony join in, for "community" grooming is a general practice among bats. Death occurs for all that have licked the anticoagulant, which causes internal bleeding.

In spite of such extermination methods, biologists say that vampire bats will not be totally destroyed. Some will continue to live in the jungles, finding victims among wild animals as they did before large ranches were established in their territory.

Besides any real reason for worry about bats, the appearance of the animals is terrifying to some people. Among those with frightening looks are certain species known as "false vampires" because, when first known, they were believed—mistakenly—to feed on blood.

One of these, living in the Amazon region of South America, is called the great false vampire because of its size. Its wingspread is thirty inches, and the head and body measure five or six inches in length. Its jaws are massive and on the tip of its nose is a spear-shaped appendage (the so-called leaf). Its ears are large and leathery. But it kills and eats small rodents, birds, and other bats. It never bothers people. Other false vampire bats live in the Far East.

Africa has an especially grotesque bat in the "hammer-headed" species. Its face appears to be greatly swollen, and there are a number of wart-like appendages on its lips, which are surrounded by ruffles of skin. With this head goes a large body, and a wingspread of forty inches!

The biggest bat on earth is the flying fox, which was given this name because its long and slender face is definitely fox-like. The creature's size is startling. Its wingspread may be as much as five feet; its body, twelve inches long.

These big bats live in tropical Asia and parts of Australia, on Madagascar, and islands of the South Pacific. Their food is chiefly fruit and berries, and a group of them raiding an orchard can do considerable damage. In some areas they make regular migrations, timed by the ripening of fruit.

With these fearsome teeth, the "short-faced" bat, native to South America, is able to eat small birds, insects, and mammals. Like many bats of this area, it has a spear-shaped "nose leaf," which is made up of skin and muscle.

However, it is likely that more people have been frightened by little brown bats, perhaps no more than two inches long, than by the giants. These are the kind that like to settle near human habitations and are apt to fly close to a person's head. They apparently have no fear of people. But the same can-

Little brown bats (a cluster of them are shown here as they hibernate in a cave) are found in almost all regions of the earth. In the Carlsbad Caverns of New Mexico, more than a million of them may be sheltered at one time.

not be said of people's feelings for them! It is an unusual person who keeps calm when a bat swoops near.

Little brown bats make up in numbers for what they lack in size. They live in almost all parts of the earth. In North America they are found as far north as trees are plentiful, and as far south as Mexico.

There are about eighty different kinds. They vary in color as well as size. In southern Asia and South Pacific islands there are species with bright orange or reddish bodies and black wings.

All species of the little brown bat live in colonies —some small, some huge. Probably the most famous of their gathering places is the Carlsbad Caverns of New Mexico, where a million or more use the great cave as a shelter. Their flight from the dark interior at each sundown is a spectacular sight.

On a lesser scale, little brown bats roost in hollow trees, rock crevices, and in barns or other "out" buildings. Several dozen may crawl into one tiny crevice. Very quiet during the day, late in the afternoon they start to squeak and move around. As soon as the sun sets, their flight begins, and soon their daytime roost is emptied as they go in search of insects. Their appetite is big. One devours enough insects between dusk and daylight to equal its own weight. The bats of the Carlsbad Caverns do away with several tons of insects every night.

Another small bat of North America is the flittermouse or pipistrelle. There is also a species of red bat, with orange-red males and chestnut-colored females. This bat has different roosting habits from the cave dwellers. For its resting hours it stays out in the open, hanging on a tree branch, and looking much like a colored leaf. A person passing by may be alarmed when a "leaf" suddenly extends wings and flies off into the sunset.

Red bats migrate, like many birds, leaving northern regions in the fall for warmer southerly climes. When migrating, they may travel in flocks of more than a hundred, but once settled for the season, they become solitary creatures, each keeping to itself. Part of their flight is made during daylight hours, and often they follow a coastline. Sometimes people on a ship fifty miles off shore have been startled by a group of bats landing on a deck.

North America's largest bat, with a wingspread of sixteen inches, is called hoary. The name might suggest something rather unattractive, but the hoary bat is quite beautiful. Its fur is soft and luxurious. Intermingled with its rich mahogany and brown fur are long white hairs which give the coat a frosted sheen. Like the red bat and other close relatives, this species migrates, and lives on insects. A hoary bat is big enough to capture luna moths and large beetles.

Of all mistaken notions about bats, one of the most widely believed is that, mysteriously, they make their way, flying and hunting, without being able to see. "Blind as a bat" is a popular phrase that keeps the idea alive.

Actually, most bats have eyes and can see well in bright light. A bat can also "see" in the dark, because it is supersensitive to echoing sounds. When one is flying, it gives out sounds continuously. Many of them are pitched too high for human ears to detect. However, the bat's ears pick them up as they are bounced back by nearby objects, and this serves to guide the flying mammal.

There are about two thousand kinds of bats. In

Courtesy of the American Museum of Natural History

A fruit-eating bat is shown sleeping. The fruit-eaters usually are larger than the insect-eating bats; they are also known to have better vision.

accordance with their diet, most of them fall into one of two large groups—the fruit-eaters and insect-eaters. But there are fish-eaters and meat-eaters, too, and some that draw nectar from flowers. And then there is the vampire that requires blood. Somehow, this seems to win more attention than any of the others!

6

Snake Fright and Snakebite

Some people would rather meet a lion face-to-face than see a snake creeping toward them. They understand lions. Without knowing exactly why, they are terrified of snakes.

Of course, if the meeting takes place where poisonous snakes are known to exist, they have reason to be concerned. The bite of a venomous snake is not to be taken lightly. But superstition and misinformation cause many a person to regard *any* snake, poisonous or not, with feelings close to terror. Will he be hypnotized by this serpent? Will it spring on him from the ground and wind itself around him? If it bites him, will he face certain death?

Knowledge is needed to deal sensibly with "snake fright" as well as snakebite because there are many different kinds, poisonous and nonpoisonous. They live in a variety of habitats—from jungles to deserts, from swamplands to mountain slopes.

It is difficult for those who fear snakes to believe there are others who admire and enjoy them. But there are—usually people who have an understanding of the reptiles. Anyone who dislikes them should learn about their habits and backgrounds to be able to separate fact from fiction. Certainly residents of country and suburban areas, and all campers are wise to find out whether there are snakes in the vicinity and, if so, whether they are poisonous.

In the United States, the poisonous snakes fall into four groups: rattlesnakes, found in nearly every one of the continental United States; copperheads, which live from New England to Texas and in all southern states; water moccasins, inhabiting chiefly southeastern and other southern states; coral snakes, native only to the deep south and southwest, from North Carolina to Texas and parts of Arizona.

Although poisonous snakes are so widespread, they are not unavoidable. In many large areas of the country there are only harmless snakes.

The poisonous species are equipped with hollow, sharp fangs that act like hypodermic needles. When they pierce a victim's flesh, venom from nearby glands is forced through them by the action of certain muscles in the head. The fangs are shed periodically, new ones growing even while the old ones are in use.

Distinguishing poisonous from nonpoisonous snakes is not always easy. For example, the venomous coral snake has bright red, yellow, and black

By careful handling, a reptile expert is able to draw back the sheaths that cover the fangs of a rattlesnake. These fangs, which inject poison into a victim, are folded back when not in use. They are positioned in the mouth to make a strike.

rings around its body. Other, harmless species, that live in the same general vicinity, are marked in the same way. Though the arrangement of the rings differs somewhat, it is not enough of a contrast to be easily noted—and surely not in an instant. In Central America, the coral and some nonpoisonous snakes resemble each other exactly.

Rattlesnakes are easy to recognize because they carry their namesake on their tails. There are a number of mistaken notions about this "rattle." For instance, it is believed the age of a snake can be determined by the number of segments in the rattle. The fact is, each segment actually is a piece of skin that fails to come off every time the snake sheds.

Since during its early life a rattler sheds its skin two or three times each year, a count of rattles on a young snake does not give an accurate age estimate. With older snakes, as the rattle grows, the end segments tend to break off. Again, a count proves nothing.

An average number of rattles found on older snakes is eight. But reports have given individuals twenty-five, or even fifty! In captivity, where the snakes are well cared for and have no encounters with enemies, they may hold on to more rattles than their relatives in the wilds.

A series of eight segments creates a better sound than a longer series. With more than that number it becomes deadened. Actually, the sound produced when a rattler vibrates its tail is more like the dull buzz of a bumblebee. Other kinds of snakes also vibrate their tails, and if one happens to be lying among dry leaves, you might think you were hearing a rattler when, in fact, it was a copperhead—or a nonvenomous species.

Apparently the reason for the rattle is to warn a possible enemy to keep away. Rattlesnakes were developing millions of years ago, in the western part of what is now the United States, at a time when a variety of hoofed mammals were flourishing there. A big hoof tramping on a snake, of course, could kill the reptile. But the snake occasionally could inflict a poisonous bite—enough to cause suffering. Gradually the rattle must have said to wary animals: "Don't tread on me."

There are twenty-nine different species of rattlers. The largest of all are the diamondbacks. In the western United States, they reach a length of seven feet; in the east, they are still larger. Diamond-shaped markings, formed by bright borders around dark blotches, are responsible for the common name.

The timber rattlesnake has dark, angular bands across the back. This was the first kind of rattler to make trouble for English Colonists when they reached New England and Virginia. It is still found through eastern and southeastern states.

To the west, the prairie rattler is widespread. It has several color variations, including black, pinkish, and "faded." The sidewinder, also common to the prairie, is a small rattler—thirty inches long at the most. It is especially well known because of the way it travels over loose sand—coiling its body into loops and throwing the loops forward.

Smallest of the rattlers is the pigmy, native to the southeastern United States and parts of the midwest. It rarely reaches a length of twenty-four inches.

Small or large, however, a rattler bite must be taken seriously. The poison injected by a nine-inch pigmy into the finger of a naturalist gave him a week of utter misery. And the finger could not be used properly for a year and a half. The naturalist had not immediately sought treatment after the bite because the snake was so small!

Bites by larger rattlers leave no doubt that help is needed. Beside the pain, a swelling and discoloring of the skin near the bite begins. Faintness, nausea, and other frightening symptoms may follow. Nevertheless, in spite of the very real danger, most victims of rattlers and other poisonous snakebites in the United States do recover.

Snakebite kits are available for people who go camping or exploring in venomous-snake country. But even without one, a person can help himself by using primitive first-aid treatment such as early Colonists learned from American Indians.

A tourniquet (a strip of cloth torn from a shirt will serve) is applied about two inches above the bite—which usually is on a leg or arm—and made tight, but not too tight. There should be just enough slack to be able to force a finger between the tourniquet and the skin. Blood is then sucked from the wound for at least half an hour. The person who is administering this aid should repeatedly spit the blood from his mouth; however, if some is accidentally swallowed, it will do no harm, as the venom goes into the stomach, not the bloodstream.

Some people recommend that, if a pocketknife is available, the victim's flesh should be cut where the snake's fangs entered. The purpose of this would be to increase the flow of blood. However, unless handled expertly, infection may result from cutting and cause additional trouble. As a rule it is better to wait for skilled treatment, if it can be obtained within an hour or so. If cutting *is* done at the scene, the knife and skin should first be sterilized with iodine or alcohol.

A victim's movements should be slow and easy. He certainly should not run, nor should he take any stimulating drink. These actions, by increasing the circulation of the blood, cause the venom to be absorbed into his system more quickly. A doctor usually can be reached within an hour, and if he is not familiar with snakebite problems, he can phone a hospital for advice.

Snakebite kits (found in sporting goods stores or drugstores in areas where poisonous snakes are frequently encountered) vary in the equipment they supply. Important items usually included are: rubber tubing to serve as a tourniquet, a knife, iodine or alcohol, a suction device to suck out blood. Some kits contain antivenin, with instructions as to its use. However, a person may be allergic to this serum, and tests must be made, as instructed on the kit, before using it.

In the United States, many more bites are reported from rattlers than from any other kind of snake. A bite is most likely to occur when a person reaches into a rock pile or a hollow space at the root of a tree or other hole, or if he steps across a log without checking the other side.

The colorings and markings on a copperhead create an excellent camouflage. The snake is hard to see among the colored leaves of autumn.

Courtesy of the American Museum of Natural History

The copperhead is of especial concern in north-eastern woodlands, from New England to the mid-west, because the pink, brown, and gray colorings blend with the fallen leaves on the forest floor, and a perfect camouflage results. Usually the top of this snake's head is copper color or bronze. There are hourglass-shaped marks on its back.

The venom of a copperhead is painful and harm-ful to a human, but it is not considered as dangerous as that of rattlers. In North America, deaths from a copperhead bite are few and far between.

A close relative of the copperhead, the water moccasin, is found within the same range inhabited by southern copperheads. As its name suggests, it fa-vors streams and lakes, although at times it travels some distance from water to spend part of its time on land. It is also inclined to hang by its tail from tree branches.

Water moccasins are larger than copperheads. They reach a length of nearly five feet and may have a great girth. While young, their coloring is bright, and their scales are patterned somewhat like a young copperhead. When adult, their color is dark brown or black. The inside of the mouth is white, and because of this, they are often called cotton-mouth moccasins.

The venom of water moccasins is potent and people are inclined to be especially terrified of them. However, while work is being done in swamplands where they exist, the men on the job do not seem to be concerned.

Courtesy of the American Museum of Natural History

Water moccasins do not always remain in
the water. They may wander on dry land or
hang by the tail from tree branches.

"Leave 'em alone and they'll leave you alone,"
they say.

This is just one piece of evidence (there are
many) that snakes are not aggressive. One of the
many false beliefs about them is that they have no
fear of people and are always on the lookout for
human victims.

The truth is that most snakes, if given the
chance, scurry away and hide rather than trying to
attack people.

But the giant venomous snakes of some countries really do belong to horror stories.

Australia has such monsters as the tiger snake and taipan. Their venom is so potent a human victim may die within a few minutes after being bitten.

In Africa the mamba is the most dreaded of all. It has a potent venom and is large—one species grows to a length of fourteen feet. It is swift-moving and more aggressive than most snakes.

Mambas are closely related to cobras of Asia, a species of snake responsible for the deaths of more than twenty-five thousand people a year. The king cobra is the largest poisonous snake in the world, measuring eighteen feet or more in length. However, the Asiatic hooded cobra is said to cause more deaths than any other species. Its venom is highly toxic and it lives in areas densely inhabited by humans.

Cobras are the most generally known of Old World snakes because of their use in show business by "snake charmers" and because visitors to zoos recognize a cobra by the way it spreads its hood. It is not the only snake with this ability (some harmless snakes, such as the hognose, do the same thing), but the cobra's hood is the most spectacular. Asiatic cobras give an especially startling effect when the hood is spread because markings on the scales make it seem that there are eyes at the rear of the head. Large, staring eyes! The real eyes are in a normal position, but when the hood is open they are not easily seen.

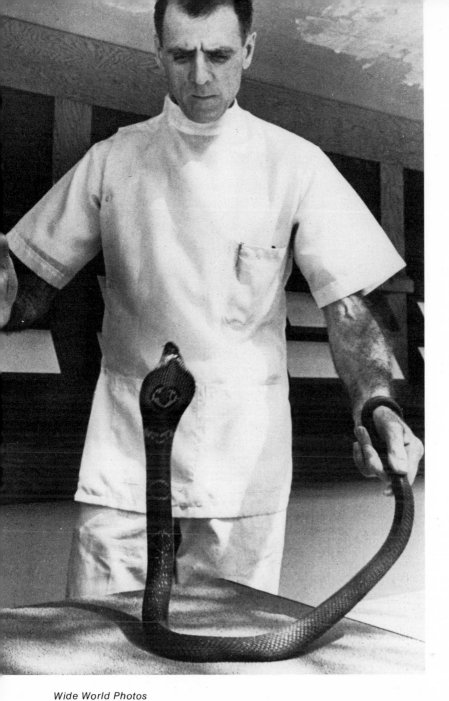

This cobra is about to be "milked" of its venom, which is an important item in medical research.

The forming of a hood is made possible by a series of small ribs on the sides of the vertebrae of the neck. When the snake becomes excited, it uses powerful muscles to pull these ribs forward, thus stretching the skin and forcing the scales wide apart.

Besides the hood, another frightening aspect of cobras is the way they pose when approached by a possible enemy—the front third of the body is raised straight up, while the rest remains in a flat coil. Because of this habit, the cobra became a perfect showpiece for snake charmers. A snake, kept in a covered basket, will rear up and spread its hood as soon as the cover is removed. The "charmer" then begins to play on a flute-like instrument, swaying his body as he does so. The cobra sways too, and it appears to be dancing to the music. However, the animal is actually deaf; its moves back and forth are following the motion of the snake charmer's body.

Sometimes performing cobras have their fangs removed, but not always. A showman sometimes chooses to live dangerously; sometimes he has received repeated inoculations of venom which render him immune to the snake's bite.

Besides the species found in India, there are a variety of cobras in Africa and the East Indies. One of them is the Egyptian asp. In ancient times, this snake was offered to political prisoners who had been sentenced to death. Its bite was presumably a sure and not-too-unpleasant way to carry out the sentence.

No African cobra has so wide-spreading a hood

as the Indian species. There are several, however, that have the power to spit their venom quite a distance. They aim at the eyes of any large animal that approaches them and show great accuracy in hitting their target. A new captive in a zoo is likely to cover the glass part of its cage with venom as it spits at visitors who come close.

Snake venom has long puzzled and fascinated scientists. Today many facts are understood, but there are still some mysteries. How the venom is manufactured is still not clear. And the venom still defies a definitive chemical analysis.

We find, though, that many mistaken notions, once widely believed, are now recognized as being untrue. No longer do people in general think that some snakes have a stinger on the top of their tail, or that snakes can sting with their tongues. And it is known that in North America no native snakes will jump at a victim, though one striking from the top of a log or rock may *seem* to do so. No North American snake can strike to a distance greater than half the length of its body.

Small and venomous, large but nonpoisonous— which kind of snake is most frightening? The answer probably varies with different people. Some may feel that the killing of prey by constriction is more gruesome than poison. Others see it the other way.

Not all constricting snakes are giants, and none of the really big ones live in North America. They are found in South America, Africa, and Asia. The snakes properly called boa constrictors may grow to fifteen feet or more, but a more usual size is a ten-foot length. Several kinds of python are much larger. And the anaconda (which is actually a member of the boa family) is the real giant. It is known to achieve twenty-five and thirty-foot lengths. Reports of still greater size—up to forty feet—have not been proven.

There are small boas too, some of them living in the United States, but only because, years ago, their ancestors were "stowaways" from South America, hidden in huge bunches of bananas.

The giant constricting snakes belong to one of two groups—the boas and the pythons. With a constrictor, an attack is made by striking at the prey, jaws open. As the teeth are sunk into the victim, coils of the body are thrown forward to envelop it. Then two or three coils are tightened so that the captive animal cannot breathe. It is not true that a constrictor crushes most of the bones of its victim. When the prey becomes quiet, the snake releases its coils and consumes its meal.

The size of the prey depends very much on the size of the snake. A ten-foot boa is well fed on monkeys, rats, and other small mammals. But a full-grown anaconda or python goes after bigger game —perhaps a tapir or small zebra. After a satisfying feast, the reptile may be quiet for two or three months, with no need for additional nourishment. Neither man nor beast would have reason to fear the

A boa constrictor makes an interesting and friendly
pet—for people who like snake pets. The father
of this young boy bought the six-foot-long reptile
for himself. When the boy and snake became play-
mates, the play was always closely supervised.

giant at such a time. It would not exert itself to attack anything.

Smaller constrictors seek such prey as mice, rats, and rabbits. Some nonpoisonous snakes, such as whip snakes, do not constrict, but pin their prey to the ground with part of the body. At the same time, the head is brought close enough to the victim so that the jaws can go to work. Such snakes also eat small rodents, birds, frogs, and other snakes.

Two types of snakes common in North America and the Old World are known as "rat" snakes and "chicken" snakes. The popular names refer to their choice of food. They kill all their prey by constriction.

One of the strange examples of death by constriction is the way a king snake kills other snakes. Almost in one motion the "king" strikes its victim and twines itself around the body so similar to its own. Then its coils press until the victim suffocates. It is the same operation carried out by giant boas and pythons against mammals.

A number of snakes include other snakes in their diets, and the snakes they eat may be venomous or nonvenomous. Those that eat poisonous species suffer no ill effects from the venom if they happen to be bitten; they have an inbuilt immunity.

A king snake battling with a rattler is sure to be the winner, even if the rattler is larger. But the king snake would not attempt to swallow so large a meal.

King snakes have a wide variation in size, some

kinds reaching a length of over five feet, others not growing as much as a foot and a half. They are found across the North American continent and may not be quickly recognized because of their various colors and patterns as well as different sizes.

The hognose snake is another species not easily recognized. It is a mimic of exceptional talent, and can imitate the actions of venomous snakes so well it is often mistaken for one. If disturbed, it will raise the front of its body and flatten its head to form a cobra-like hood. It also takes in air and then expels it with a sharp, hissing sound. It may also open its mouth and strike out with the technique of a rattler.

We may well expect trouble from such a serpent. Yet the hognose (sometimes called puff adder, spreading adder, or blowing viper) is entirely harmless. It feeds on toads and other amphibians and small reptiles. It may grow as long as four feet and has a stout body with dark coloration. One kind is often mistaken for a venomous water moccasin.

The hognose is a favorite with people who enjoy snake pets. It is easy to feed and grows very tame with proper handling. Boa constrictors also are praised by snake fanciers as being interesting and friendly in captivity.

Obviously the story of snakes is not simple. A kind of animal that can be someone's treasured pet, and also be the reason for a person's death or suffering, is complicated. There are snakes and snakes and snakes.

7

Survivors from the Age of Reptiles: Alligators and Crocodiles

In her *Adventures in Wonderland,* one of the strange verses Alice recites has to do with a crocodile. It concludes:

> How cheerfully he seems to grin
> How neatly spreads his claws,
> And welcomes little fishes in,
> With gently smiling jaws.

Not many people think of a crocodile as smiling gently! They look at the reptile's huge mouth, filled with long pointed teeth, and shudder. Here is a real monster, they think. A sure killer! A man-eater!

Actually neither of the conflicting opinions is entirely fair. Crocodiles are certainly not gentle. But neither are all of them man-eaters. They kill animals for food—as do all flesh-eating creatures. But in some areas of Africa, children swim and play in rivers inhabited by crocodiles without being bothered by them. However, at other villages nearby, the same species of crocodile is an ever-present danger to those who go to the river for water. No single description fits them all.

Crocodiles, and alligators, too, change their diet during the course of their lives. When very young, they eat quantities of insects and shellfish. Somewhat later they feed largely on fish. When adult, their prey includes mammals that come close to the water, birds—and many young crocodiles.

The size of the prey depends on the size of the crocodile. A big one, perhaps a fifteen-foot individual, searching for food along an African riverbank, may grab an antelope or deer. It is a very large and old crocodile that is most likely to tackle a human. Its method of dealing with a large victim is to pull it beneath the water's surface, then roll and twist the body while tearing it to pieces.

Its teeth are designed for tearing—not biting or chewing. Therefore, the food is swallowed in huge gulps. Digestion might not be possible except that the crocodile instinctively swallows quantities of stones, and they remain in the stomach, acting as grinders to crush bulky pieces.

Teeth make a noticeable distinction between crocodiles and alligators. In an alligator, the teeth of the upper jaw are placed outside those on the lower jaw. And two enlarged lower teeth fit into pits in the upper jaw, so that they are completely out of sight when the mouth is closed.

Gordon S. Smith from National Audubon Society

An African crocodile displays its sharp teeth, designed for tearing, as it rests with its mouth open. Although a crocodile's jaws have strong hinges, they do not open easily; most of their power is in the closing mechanism.

With a crocodile, the teeth of upper and lower jaws are pretty much in line. And each enlarged tooth on the lower jaw fits only into a niche, rather than a pit in the upper. They can be plainly seen when the jaws are closed.

Another distinction is in the shape of the head. With the crocodile, the snout tapers from a considerable width on the back of the jaws to a narrow snout. In the alligator, the sides of the jaws are almost parallel, and the snout is bluntly rounded. The crocodile's head is roughly triangle-shaped; the alligator's is a rectangle.

An interesting bit of nature lore concerns the crocodile's teeth: When the giant reptiles are basking in the sun on the banks of the Nile River with mouths open, large birds go from one to another, apparently picking at their teeth. They run in and out of the fearsome mouths—and the crocodiles make no move to swallow them. Apparently the free dental service is appreciated.

Because of the way crocodiles rest, with the lower jaw flat on the sand or earth and the upper raised, a belief grew that they never moved the lower jaw. But, in fact, the lower jaw is hinged to the skull as in other animals and functions along with the upper. A resting crocodile opens its mouth by raising the rest of the head and bringing the upper jaw along with it!

Though the jaws are built on strong hinges, most of the power lies in the closing mechanism. They cannot be opened so easily.

The tail is actually as dangerous a weapon as the jaws.

Crocodiles are so well designed for survival, it is not surprising that they have been on earth for millions of years—since the Age of Dinosaurs. They are at home both on land and in water, where they can be submerged, yet able to breathe and see. This is because of their extraordinary nostrils, eyes, and throat.

The nostrils, at the tip of the snout, are raised above the flattened head, so they can take in air when the body is under water. When submerged, the nostrils are closed by a valve, and a fold of skin shuts off the mouth cavity from the windpipe so that as the crocodile grasps its prey below the surface, its lungs do not fill with water.

The bigger the crocodile the more alarming it looks. And so the award for the "most frightening" must go to an Indian species, the gavial. Its adult length averages between twelve and fifteen feet, but some specimens attain twenty feet or more.

There are records of trinkets such as bracelets and rings being found in the stomachs of these great monsters. However, it is believed they were there because the crocodiles had eaten corpses floating on the Ganges River; the gavials are not known to attack living humans.

Besides such gigantic forms, there are medium-sized species, as found in New Guinea, with an average eight-foot length, and dwarf species, like the Congo "dwarf," which rarely becomes more than three and a half feet.

Crocodiles, like other reptiles, continue to grow throughout their entire lives. However, after the first few years, the rate of growth slackens, and in later years is scarcely noticeable. Their normal life span varies according to the species. Some African species have been known to live more than twenty years. Color of skin and shape and size of the body are all factors in estimating the age of an individual.

It appears that alligators live longer than crocodiles. American alligators in captivity often thrive for more than fifty years.

Such are the records kept by scientists. But in various parts of the world, visitors to reptile show places may be told that certain crocodiles or alligators on display are hundreds of years old. Or per-

Courtesy of the American Museum of Natural History

On land, alligators can run with considerable speed—but for short distances only. They usually lie on their bellies, but on occasion raise themselves on their legs in what has been called "a curiously dinosaurlike pose."

An alligator's nostrils are raised on the tip
of the snout and are able to take in air when
the rest of the alligator's head is submerged.

haps a thousand! This is one way "show business"
contributes to general misinformation.

Until about fifty years ago, American alligators
flourished in waterways from Louisiana through the
Gulf states and as far north as North Carolina. The
most dense alligator population was in Florida,
where the animals swarmed in nearly every stream
and lake. They doubtless were alarming to people
who wanted to clear the land and build communi-
ties, but not to professional hunters. Alligator skins

were in demand for such apparel as shoes, purses, and belts. Instead of a gold rush, an "alligator rush" developed, and the slaughter was incredible. One family engaged in the business were proud of their record. Twenty thousand victims in one year! Other hunters were not far behind.

When killing alligators for their skins was legal, hunters worked at their trade day and night. In the daylight they used barbed hooks to stab their victims in the back, then pull them from their hide-outs. At night hunting was even easier. The men would glide quietly in a boat through open areas of a swamp, then pierce the darkness with a high-beam light. Whenever they saw two shining eyes, they would fire a rifle. The target was not difficult to hit: the bullet would be sure to enter the alligator's head.

Meanwhile, collectors for zoos and reptile farms were busy capturing live specimens. It was quite possible to collect sixty or seventy in a single night.

Ross Allen, when founding his famous Reptile Institute in Silver Springs, Florida, enjoyed the ex-citement of catching alligators in his bare hands, then bringing them to the Institute unharmed. One method was to swim after an alligator and, catching up with it, throw his leg over the body and get a scis-sors hold on it, while wrapping a rope several times around the snout.

Another approach was to have a noose at the end of a long pole. With a companion, Mr. Allen went out in a canoe, paddling until an alligator was seen. While his companion continued to paddle, steering alongside the animal, he floated the noose over the alligator's head or throat, then held it while he tied the jaws.

Catching crocodiles by these methods would be almost impossible, since they are more alert and swift than alligators, especially when in the water, and more aggressive. In captivity, also, differences in the nature of the two kinds of reptile are noticeable. Alligators are more easily handled and dependable in their actions. Crocodiles often are vicious, and even after seeming to have settled down to life in captivity, may charge ferociously at a keeper for no known reason.

There is something eerie about the eyes of an alligator or crocodile at night. Although green in daylight, after dark, in the glare of a flashlight or a boat's spotlight, they gleam red or pink.

Relentless killing of alligators continued until —certainly in the United States—it seemed the animals would soon be extinct. But today conservation laws are giving them a chance to hold their own. In some areas their numbers are definitely increasing. And now and again in the South one may startle people by appearing in a watery part of a suburban golf course or in a private swimming pool. One may even commit the crime of snatching an unlucky pet dog. In which case, alligators of that community are quickly returned to the category of "frightening monsters."

8

Terrors of the Sea: Sharks, Octopi, and Squids

Frightened of a fish? Yes; when the fish is a shark, we may have good reason to be. As with snakes, however, many species have an undeserved reputation as man-killers—fierce "enemies" that attack because of a sinister delight in destroying human life.

But there are some sharks that do not harm anything larger than shrimp, crabs, and other small invertebrates. The whale shark, largest of all, is one of these. Other species may attack humans when threatened or disturbed, but are not notable man-killers. Still others are more ferocious and, without provocation, will attack humans or even small boats.

There are about three hundred different kinds of sharks. Some grow only a foot long, some reach a length of forty feet or more. They vary not only in size, but in form and habits, and in the development of their greatly feared teeth. In some, these consist of a single row; others have two or more rows set at different angles. In many, the teeth tip forward when the jaws open and backward when they close.

Shark teeth are not fastened firmly in the jaws. They are merely set in the gums so that they often fall, or are knocked, out of the mouth. This might seem a serious problem, but a shark is not handicapped for long. There are always "replacement" teeth behind those that are lost and they quickly move forward, ready for action.

Sharks live in ocean waters throughout most of the world, except where it is extremely cold. Some species also go into fresh water, swimming along rivers and occasionally in lakes.

However, it is not the number of species nor where they live that most concerns people. The big questions are: At what beaches are they likely to attack? How can swimmers, fishermen, or victims of ship or plane wrecks avoid them?

Many answers are available, thanks to continuing research. Several countries have a shark research panel. In the United States this is a part of the American Institute of Biological Sciences. Members of these panels encourage people anywhere to report instances of shark attacks, and press-clipping services are employed to send in published accounts. A doctor or scientist of the region is then asked to check out the story, making sure that all the evidence is reliable. The research panels keep a record of the facts that emerge, and every so often hold conferences to discuss findings—old and new.

It is evident, they report, that attacks can occur at almost any time or place where there is a person in the water. But they most often occur during the summer in temperate regions, and year round in the tropics.

In the United States attacks have been reported in coastal waters from Massachusetts to Florida, and on the west coast as far north as Trinidad Bay, California. The majority have been from the Florida area. After Florida, on the east coast, New Jersey and South Carolina have the most recorded attacks.

Over a four-year period there were recorded— worldwide—attacks on a hundred and sixty-one swimmers, waders, and others participating in water sports at or near the shore, who had done nothing to provoke the animals. Also there were a number of other attacks on people who had been trying to spear sharks or otherwise disturb them.

In the same period of time, where ship sinkings or plane crashes hurled helpless bodies into the sea, at least four hundred and seventy-six attacks were made. "At least" is the way to report this figure, because there probably were a number of attacks that went unreported.

Of all countries, Australia has the most trouble with sharks. The United States is in second place for this unwelcome honor. While many attacks are made a considerable distance out from shore, they occur, too, in harbors, creeks, and rivers. Sometimes they are in very shallow water, no more than knee-deep.

Beaches, especially long, open ones, have been the setting for many attacks. Sharks are attracted by vibrations although, as a rule, they seem to stay away from areas where hundreds of people are making a commotion. Experiments have shown that sharks which do not react to the smell of blood quickly respond to vibrations. Motion in the water *and* blood produced a truly furious reaction.

Many divers in South Sea waters are aware of this, and if a shark is noticed, they try to remain absolutely still.

Although horror stories (both fiction and fact) are inclined to have sharks doing their deadly work mostly at night, investigations contradict such timing. It seems the frightening fish are most apt to strike in the late afternoon. Thor Heyerdahl, who made a journey of over four thousand miles across the Pacific on a raft, and described his experiences in the book *Kon-Tiki,* noted that the sharks he saw always began to hunt towards dusk, but were not so active after dark. In a number of places, because of sharks, from three to six in the afternoon is considered the most dangerous time for swimming.

Over the years, many efforts have been made to find means of protecting unlucky people from the fierce sea-hunters. During World War II, scientists in the United States Navy worked to create a shark repellent which they could issue to sailors and airmen likely to be downed in the Pacific. Some "shark chasers" were invented, such as a cake of dye and chemicals enclosed in a waterproof pouch. Where this was released, it sometimes discouraged attack for several hours. But this and other inventions could not be depended on; there was always an element of doubt about their success.

It might seem an easy matter to keep beaches free of sharks. Build a barrier! A rigid, permanent fence is one type. But it must reach the bottom of the water for the entire length of the beach and be high enough so that even abnormally high tides will not go over it. Its slats must be close enough together to prevent small sharks from penetrating it. Such a barrier is costly to construct and maintain, however, and since shark attacks are few and far between, it usually is considered an unwarranted expense.

A practical plan for protecting open beaches is "meshing," which was originated in Sydney, Australia, more than thirty years ago. With this system, huge flat nets are suspended, curtain-style, in the water. The meshes are large enough to permit the head—but not the body—of a shark to pass through, and in this way it is trapped. Entangled bodies are later disposed of on shore.

Any swimmer who encounters a shark is not likely to look it over calmly to decide what kind it is. Nevertheless, it is worthwhile to know the characteristics of the most frightening species.

The great white shark has a second name: man-eater. It is well deserved. This is the most ferocious of all sharks, and is dreaded because of its attacks on people and boats.

Individuals frequently reach a length of twenty feet or more; the largest on record measured over thirty-six feet. The great white shark's body is heavy and thickset. Its teeth are relatively few, but those of the upper jaw are large and triangular, with serrate edges. Despite its name, the only white part of the body is its underside, and this is a feature of other sharks as well. Its back may be pale gray, dark gray, or even black or brown.

The white shark swims in all temperate and warm seas, but is known to roam into colder waters. Though it definitely belongs to the deep ocean, it

The whale shark, largest of all sharks, is not a man-eater. Instead it lives on small animals such as shrimp and crab.

Courtesy of the American Museum of Natural History

Courtesy of the American Museum of Natural History

The great white shark is known as the "man-eater" —a name earned because of the many people who have become its victims. Man-eaters have been measured at a length of more than twenty feet.

may prowl along coastal areas and even enter the mouth of a river. It regularly preys on sea lions, seals, turtles, and a variety of fishes. Man is sometimes a victim.

There is occasionally a conflict of opinion about the name "tiger" shark. Some people claim it stems from the tiger-like pattern on the skin of the young (the colors fade as the shark matures); others believe "tiger" refers to the shark's hunting instincts. Its prey includes practically anything that lives in the sea, from snails to giant sea turtles and other sharks. But it is also a scavenger and eats dead bodies or any kind of refuse that comes its way.

Tiger sharks are found far out at sea and close to shore in the tropics. During summer months, they go farther north. Some have been seen off the coast of Massachusetts. They do not have a really bad reputation for attacking humans, but there have been cases of death or injuries from shark bite where a "tiger" was known to be responsible.

The blue shark comes by its name honestly; the upper parts of the body are a brilliant blue. The underside is white. This species lives in both tropical and temperate seas, and it has been noted that in the warmer regions the "blues" stay deep in the water while in cooler areas they swim at its surface. Mostly they eat fish, but the carcasses of whales are also a favorite food. They are known occasionally to attack men and boats.

Another shark with a blue back is the mako. This is a species that is often called savage, but perhaps the description is somewhat unfair. The savagery for which it is famed results from its pursuit by game fishermen.

The blue shark is known to occasionally attack men and boats. It can be found in both temperate and tropical seas.

Courtesy of the American Museum of Natural History

The fishermen are challenged by the way mako sharks will take fast-moving bait, then swim rapidly with it, leaping out of the water, perhaps as high as ten feet, in attempts to escape. Sometimes, then, they deliberately and fiercely attack man and his boat in their efforts. But the encounter has not been of their choosing.

Makos may reach a length of twelve or thirteen feet and weigh up to a thousand pounds.

The hammerhead shark has one of the strangest heads to be found on any animal. Its skull is flattened into two long, narrow projections, with eyes and nostrils located at the extreme ends. There are a number of species and they vary in size, some growing to twenty feet in length. Attacks on humans by hammerheads of all species have been reported, from California, Florida, and other localities.

Other kinds of sharks are known to kill and devour people on occasion. But even experts aren't always certain of how accurate reports are. For instance, until recently, the gray nurse shark was believed to be the most dangerous of all in Australian waters. Now researchers say that cases of mistaken identity gave it that bad name—undeservedly.

There is still much to be learned about sharks. They are strange fish. In fact some scientists object to their being classed with fish, because of certain differences in their makeup. A shark's skeleton is composed of gristle instead of true bone. Other fish have a gas-filled swim bladder, which is related to the lungs, and which makes it possible for them to hover, weightless, in the water. Sharks are not so equipped; they must either keep moving or sink. There are differences in the gills, the type of scales, and methods of reproduction in bony fish and sharks.

Through many centuries these "gristle" fish have commanded man's attention—in fear and in respect, too. Years ago the Maoris of New Zealand made gods of sharks. Certain West African coastal tribes worshiped them. In India fantastic rites were held, where people deliberately presented themselves to hungry sharks to be devoured.

The octopus has never given humans anything like the trouble sharks have. But it has the reputation of being a sinister, man-eating monster. Indeed, it seems impossible for this weird creature to live down stories concerning it that were written more than a hundred years ago and have been popular ever since. For one, Victor Hugo in *Toilers of the Sea* had his hero attacked by an octopus in an underwater grotto. It was a huge specimen that began by seizing the man's right arm, and soon five of its eight arms were fastened by suction to his body while the other three adhered to the rocky grotto wall.

The battle between man and beast that followed is horrifying, and so are the descriptions the author gives of the animal as "glue filled with hatred," "phantom as well as monster," "sombre demon of the water," and "Medusa [the mythical Greek

The octopus has been a favorite "villain" with
science fiction writers for many years, but
only rarely does one cause trouble for a human.

whose head was covered with snakes instead of hair] served by eight serpents."

Small wonder the octopus has been looked upon as a fiend, to be avoided at all costs by anyone venturing at sea. And because of its odd appearance and power of holding by suction, rather than the usual grasp of paws or hands, it will probably continue to be a popular "villain" in science fiction.

Most eerie are the eight long, tapering arms, or tentacles, attached in a circle surrounding the head. They can coil and turn as if made of rubber, but they are strong and muscular. On the inner side of each are suction cups, graduated in size. Large specimens may have as many as a hundred and twenty pairs of these suckers on each arm.

It is rather gruesome to realize that a mouth is in the center of the spreading arms and that the prey is carried to it. The octopus then pumps a digestive juice over the victim and slowly sucks at the softened flesh.

There are species of octopi that grow as big as twenty-eight or more feet across the outstretched arms. These could be terrible adversaries for any unfortunate human that became entangled in the giant arms. It does happen occasionally with pearl divers or shell collectors on coral reefs of the Pacific.

Some varieties are no more than six inches long. An average octopus of Atlantic waters, fully grown, measures only four or five feet across. Crabs, mussels, and crayfish are its usual prey. It approaches such animals stealthily, then makes a swift grab. No great monster this, but it has all the unique abilities of its giant relatives.

The giant squid, though not frequently used as a fictional "frightening monster," has a gruesome appearance. Built like a torpedo, its body length may be as much as nineteen feet. Tentacles extended from the body may be thirty-five feet long, with suckers on the undersides more than an inch across. The squid uses these great, grasping arms to seize its prey and bring it close to its head. There the food is taken over by shorter arms and relayed to the mouth.

Giant squids live deep in open sea waters, and few have ever been seen alive. However, they are well known through dead bodies washed up on beaches—particularly on the shores of Newfoundland. A glimpse of one in motion could surely be the basis for some fantastic "sea serpent" story.

Smaller squids are numerous all along the Atlantic coast of North America and in the Pacific Ocean. The common species of the Atlantic is about eight inches long.

Both squids and octopi are noted for the "ink" screens they produce. When threatened by an enemy, they can, from a pear-shaped sac between the gills, shoot a brown inky fluid. This forms a dark cloud in the water, confusing the pursuer while the squid or octopus quickly darts away.

9
Birds of Prey

Bird sounds can frighten people. So can the looks of a bird. And so can just plain birds—at least in fiction. Some years ago, a horror story by Daphne du Maurier, called "The Birds," depicted our supposed "feathered friends" taking over the earth from people. They were not monstrous birds of prey—just hordes of small, medium-size, and larger birds. But all were aggressively against humans, and there were too many of them for people to combat.

This idea was dramatic for the purposes of the story. However, it was not in keeping with the facts: In truth, birds need protection from people, not the other way around. With the destruction of their breeding grounds and the use of pesticides, which often set up a damaging chain reaction in the natural world, birds have suffered great losses. A number of species are facing extinction.

The variety of birds in the world is tremendous.

More than eighty-five hundred kinds have been recognized. Among them are the beautiful, the ugly, the sweet singers, the hooters and the squawkers, the dainty sippers of honey and nectar, and the fierce hunters of mammals, reptiles, and fish. Of these numbers a few, fairly or unfairly, are generally considered to be more frightening than lovable.

Owls are among these. They are not often seen, since they do their hunting after dark. But the eerie sounds they produce are enough to make anyone not familiar with them shiver with fright. In ancient times, and to some people even now, an owl's hoot or screech was connected with superstitions, usually foretelling tragic events, especially coming death.

Screech owls are probably the most familiar of owls in North America. They nest close to people's homes, and when their melancholy cry shatters a quiet night, the effect is startling. Naturalists trying to translate it into "people language" have suggested *Oh that I had never been bor-r-r-r-n.*

Striges, an ancient Latin word for witch, also was the Roman name for a screech owl.

Screech owls are small, rarely having a body length of more than ten inches.

The voice of the great horned owl is quite different from the screech owl's cry. It is a deep, resonant hoot. *Whoo! Whoo Whoo! Whoo! Whoo!* To another owl this is pleasing; it is a mating call of a male or female. But creatures of the forest that

might become victims of the great birds, and some
people who hear it, find it terrifying. The great
horned owl has other calls as well, and may hoot
while flying as well as when perching.

This species is sometimes called the "tiger" of
the bird world. It is the most powerful and one of
the largest of all owls, having a body length of about
twenty-three inches. It has feet like grappling hooks
and can carry off a full-size turkey in its talons.

In Italy the horned owl is especially feared by
some people. When it is heard near a home, any sick

The cry of a screech owl is so eerie that
it has caused the bird to be associated with
frightening superstitions. The owl is not big;
its body length is rarely more than ten inches.

Courtesy of the American Museum of Natural History

person within is supposed to die within three days. If no one is sick, it is expected that illness will soon strike the family.

A characteristic of owls that makes them startling is in direct contrast to their hoots and cries. This is their silent flight. The usual flying bird produces a noticeable whirring noise as its wings cut through the air. But not most owls. The structure of their feathers is different, and their wings are large in proportion to their small bodies. They appear to be heavy-bodied because of their loose, fluffy plumage, but really are not. The sudden appearance—and probably disappearance—of an owl, therefore, can take place without the slightest hint of sound.

Hawks are not surrounded by the mystic atmosphere of night; they are active in broad daylight. But they are fierce hunters, in the manner of owls.

The largest of them all is the American goshawk ("gosling" hawk) which has a wingspread of almost four feet and averages about twenty-two inches in body length. Unlike other species which prey chiefly on birds, it attacks small mammals. It also will savagely attack a person who happens to come near its nest.

The goshawk is trained by some bird fanciers to hunt rabbits and game birds, in the manner of a falcon. However, a true peregrine falcon has no real rival in the field—it is enormously powerful and perfectly streamlined. Its power dives are estimated

Hawks, as may be seen by this red-tailed species, have powerful legs and ironlike claws, which are typical of birds of prey.

to be a hundred and seventy-five miles an hour.

The sharp-shinned hawk and cooper hawk are especially disturbing to farmers. They seize chickens and escape with startling speed.

All hawks seem to wear an angry expression and a threatening frown. Actually, this is the result of a peculiar strip of bone over each eye which acts as a shield against the sun as the bird soars into the sky. No other bird has this feature.

Eagles, like their close relatives the hawks, are bold, fierce hunters. But because their appearance is so majestic, they are more admired than are the smaller birds.

In the tropics, however, are certain species that look more ferocious than beautiful. The great harpy eagle is one. Its face could be used as a model for an ogre—a cruel face with a large, hooked beak. Because of its appearance, it was given the name Harpia after one of the monsters of Greek mythology. The mythical harpy was supposedly half-bird, half-woman. It was a character that snatched food away from people who had incurred the disfavor of the gods. It also was supposed to seize and carry away the souls of dead people.

Harpy eagles, of course, do nothing of the sort, but busy themselves hunting for monkeys, sloths, and other creatures of the forest. Their powerful legs and tremendous claws are unequaled by any other bird of prey.

On the Philippine Islands of Luzon and Mindanao is a huge relative—the monkey-eating eagle. This also has a bizarre appearance, with a crown of pointed feathers on its head and a narrow beak, placed very high on the face. It is not well known; it lives in such secluded areas, it is rarely seen by humans.

Vultures, on the other hand, are found in many areas of the earth. Old World vultures are especially numerous in Africa. They also live in southern Europe and Asia. In the New World, they range from southern Canada to many parts of South America.

In flight, vultures have an image of grace and beauty. But close up, they are somewhat repulsive because of the head on which, instead of feathers, there is only bare skin. This may be red, as with the "turkey" vulture, or yellow, blue, violet, black, or scarlet, as with the "king" vulture. The effect is not

The habit of "waiting for death" makes vultures unpopular with people. The birds (such as the hooded vulture shown here) do eat flesh that would otherwise lie about and decay; in this way they serve as natural aids to sanitation.

Mark Boulton from National Audubon Society

attractive. Added to their strange physical appearance, vultures have a habit that people find repulsive. For the most part they feed on carrion and they possess an uncanny ability to detect the bodies of dead animals from the air. In fact, they seem to sense when creatures below are about to die and hover impatiently, waiting to close in for a feast.

In defense of the unattractive bare head of a vulture, it must be said that the lack of feathers is important to the bird's health. Its food often contains infectious bacteria, but after it has eaten, the head which was in close contact with rotting flesh is bathed by the sun's purifying rays. If covered by feathers, this healthful treatment would be prevented.

And in defense of their constant search for dead and dying animals, we should see these birds for what they are—an excellent sanitation squad. As they devour decaying flesh, they make the countryside cleaner than it would be otherwise. Remarkably, their digestive system destroys the harmful bacteria in their food.

10

Tarantulas, Other Spiders, and Scorpions

Spiders are the smallest of all frightening animals. Like snakes, they vary from harmless kinds to those that can inflict serious injury, and even death.

The poisonous spiders are a special problem because they are so small. If we keep a sharp lookout for snakes when in poisonous-snake territory, we have a good chance of seeing them. Spiders can be completely hidden. For the most part, they are timid creatures and do nothing to make themselves noticed. But a camper may suddenly find one in a bedroll, or in clothing, packages, or papers. If his skin comes in contact with the little creature, he may be bitten before he knows what has happened.

The tarantulas, the largest and most fearsome of spiders in appearance, are not, on the whole, the most dangerous. Some kinds, native to South America, can inflict a serious, venomous bite, but North American species are likely to be no more harm-

Courtesy of the American Museum of Natural History
"Tarantula" has become a common name for large,
hairy spiders throughout the world, although
this type includes more than one spider family.
Those in the tropics, as this one from Mexico,
may have a leg spread of as much as eight inches.

ful to a human than a bee sting. In Europe, the kind
that were responsible for the name tarantula are
quite harmless.

The origin of this name is an example of how a
superstition can start, and how blame can unfairly
be attached to an animal. During medieval times in
the southern Italian city of Taranto, a large hairy

spider was quite common, and in the town the citizens enjoyed a type of frenzied dancing which the local government decided to forbid. But the dancers were not to be stopped, and when apprehended, they claimed they had been bitten by one of the spiders. They further insisted the only way to combat the poison was to dance—a wild, exciting dance.

In time, the music that had set the style for such dancing became named after the town—the tarantella. And the hairy spiders were given the name tarantula. They were then believed to be dangerous in some mysterious way.

In recent years the effect of their bite has been checked, and it was found to produce practically no ill effects.

The European tarantulas belong to the wolf spider family, which derives its scientific name, *Lycos*, from the Greek word for wolf. It is a large group, and most of its members are popularly called tarantulas. They are not web-weavers, but make retreats under logs or stones, or construct a tube-like burrow in the earth.

Many wolf spiders are active hunters, and may be found running about at night as well as in daylight hours. But some kinds spend almost their entire lives in their burrows, waiting at the opening to catch their prey.

The South American tarantulas are some of the giants of the spider world. In Brazil there is one with a body length of about three inches, and legs, when extended, having a span of nearly ten inches.

Such tarantulas really are a menace to people. They are likely to wander into a house, then settle down to rest. When an unsuspecting person comes up against one, the spider strikes. The venom it injects can inflict serious harm, but a serum has been developed to treat the poisoning successfully.

Nearly every kind of spider has poison glands and chelicerae—a pair of nippers attached to the head. In general, this equipment serves to kill or subdue their prey so it can be eaten. Each gland, located near a fang, is opened by a pore. It is covered by a layer of muscles, and these can be contorted to expel the poison. The amount ejected is controlled.

When a person is bitten, it is because the spider is frightened and acting in self-defense. With many spiders, the fangs are not strong enough to pierce human skin, so they are incapable of "biting."

Strangely, the biggest spiders do not have the most deadly venom. While the poison of tarantulas and other wolf spiders may produce no more discomfort than the sting of a wasp usually creates, the little black widow often causes serious problems.

Black widows have a sinister reputation. Their name was inspired by the fact that the female often kills—and sometimes eats—her mate, becoming responsible for her own widowed state. (This habit is practiced also by many other spiders.) Other popular names for the widow are hourglass spider, because of a red hourglass-shaped mark on the abdomen, and the shoe-button spider, because of the form of the black abdomen. These are the features generally noted, but in different areas of the United States, they vary considerably.

A fully grown black widow female is seldom more than half an inch long. The male is much smaller and is *not* venomous. Black widows live in cool, dark places and often seek out crannies in cellars and sheds or under porches, boards, and refuse. Their webs form an irregular mesh, usually built quite close to the ground. They are found in nearly every part of the United States, but are more common in the south and west than in the north.

Male (to the right) and female black widow spiders are shown here enlarged. The female is seldom more than half an inch long; the male is much smaller.

Courtesy of the American Museum of Natural History

Although the effects of a black widow's bite varies with different people, usually a victim notices his abdominal walls becoming rigid, a tightening in his chest, and a contraction of his leg muscles. There may also be a rise in body temperature, nausea, and profuse perspiration.

First-aid treatment for a bite is essentially the same as for snakebite. If possible, a tourniquet should be applied above the wound. The blood flow should be increased by suction by the mouth or a snakebite suction pump. As soon as possible, the victim should get to a doctor who may give injections of serum developed for this purpose. Other "home" treatments consist of bed rest, large quantities of water to drink, and iodine or other antiseptic applied to the bite.

For many years the black widow has been the top-ranking "dangerous spider" of North America, but recently its reputation has been overshadowed by a newcomer—the brown recluse. Its venom is extremely potent.

This spider is slightly smaller than the black widow. Its oval-shaped body is about half an inch long; its brown coloring varies from chocolate to fawn. There is a dark violin-shaped band on the head which distinguishes it from other spiders.

It was about 1957 that warnings concerning the brown recluse were first heard. Investigators at the University of Missouri had been studying a mysterious type of wound. Whoever was suffering from one was suffering also from a serious body infection. The wound itself was about three inches long and was surrounded by blisters. It would show no improvement when it was treated and, in fact, tended to develop an ulcerous condition. At last doctors found the "wound" was really a bite of a spider— one closely related to a South American species which inflicted the same damage by its bite.

The North American species then became known as the brown recluse. It was soon to be discovered in a number of midwest, southwest, and southern states.

No one knows how long the recluse had been living in the United States before it was identified as a source of serious poisoning. Like the black widow, the recluse spiders seek dark corners, inside or outside of buildings. Their habit of hiding in anything readily available results in their frequently being transported to new areas as they "hitchhike" with traveling vacationers.

Both male and female are venomous, but the female ejects twice as much poison as the male when it bites. A serious problem with the bite of a recluse is that sometimes no pain or "sting" is felt by the victim at the time it occurs. Several hours may pass before symptoms develop. As with the black widow, these include fever, nausea, abdominal cramps, and muscular pains.

Treatments have been worked out, especially at the University of Arkansas Medical Center, that can successfully combat the damage done to a human by the recluse spider. Medication may be needed for at least a week and sometimes, if the wound will not heal, a skin graft is necessary.

Scorpions are considered "relatives" of spiders because both belong to the group of animals called arthropods—animals with jointed legs. But scorpions are distinctly themselves. They have an abdomen ending in a tail-like section that is armed at the tip with a curved stinger. It arches over the back and waves in all directions.

The stinger, with its venom, delivers a fatal jab to small animals on which a scorpion feeds. To people or large animals, the poison is not fatal, but it can cause a very painful wound.

Various kinds of scorpions live in warm, dry areas around the world, and for hundreds of years they have been much talked and written about— and very much feared. Except for scientific studies, they have not been closely observed, and many strange superstitions have been created concerning them. One states that a scorpion, if trapped by an enemy with no chance of escape, will sting itself to death.

Several hundred species of scorpions are known, thirty-five being native to the United States. Most of them live in the southwest and west. In appearance they vary from shiny black, eight-inch individuals to species only an inch in length and pale in coloring.

A mother scorpion carries her young on her back. The weapon with which a scorpion attacks is in the tail—a potent stinger.

There are also "whip" scorpions that look much like true scorpions. But a whip scorpion has no stinger or poison. The worst it can do to a person is to give a sharp nip with its spiny pinchers.

The effect of venom injected into humans by spiders —or by poisonous snakes, scorpions, or insects— varies with the physical condition of the victim. Someone in poor health, elderly, or very young is likely to suffer far more than a healthy, vigorous individual. In some cases, the sting of a wasp or bee may prove fatal; in others, a presumably "deadly" bite can be overcome successfully.

As we look at the "frightening" members of the animal world, it is interesting to see that small, timid creatures rather than great monsters often present real cause for concern. Though they may not intend to harm people, their natural means of obtaining food must put us on guard.

The black widow and brown recluse spiders, rattlers and other poisonous snakes, big "wild" animals in zoos, some sharks—all have the power to inflict suffering and possibly death. But rather than condemning them, it is worthwhile to look into their backgrounds and place in the world today, and to understand their special problems in trying to survive.

Index

Abominable Snowman, the, 39–40

Adamson, Joy and George, 25

Allen, Ross, 87

Alligators, 80–88; *Illus. 85, 86*
    compared to crocodiles, 81–83, 85
    capture of, 86–88
    dog-snatching by, 88

Bats, 51–61; *Illus. 53, 57, 58, 61*
    brown bats, 57–59
    "false" vampire, 56
    "flittermouse," 59
    "flying fox," 56
    food of, 61
    fruit-eating bat, 61
    "hammer-headed" bat, 56
    hoary bat, 60
    rabid, 54
    red bat, 59–60
    seeing ability of, 60
    "short-faced" bat, 57
    vampire bat, 53

Bears, 30–40; *Illus. 33, 35, 37*
    Alaskan brown bear, 36–37
    black bear, 30, 34–35
    cub development, 38–39
    food of, 31–32
    grizzly bear, 30–33
    hugs by, 30
    Kodiak, 31, 36
    moon bear, 35–36
    myths concerning, 38–39
    performing bears, 38
    prehistoric, 31
    sun bear, 36

Birds, The, 102

Birds of prey, 102–109; *Illus. 104, 106, 108*
    eagles, 107
    hawks, 105–106
    owls, 103–105
    vultures, 107–109

Canis lupus, 7, 11

Cats, great, 18–29

Coyote, 17

Crocodiles, 80–86, 88; *Illus. 82*

compared to alligators, 81–83, 88
diet of, 81
gavial, 84
growth pattern of, 84–85
sizes of, 81
Cunningham, Alyse, 49

Darwin, Charles, 53
Dinosaurs, 1
Dracula, 52
du Maurier, Daphne, 102

Eagles, 107

Fossey, Dian, 45

Goodwin, George, 15–16
Gorillas, 41–50; *Illus. 44, 47*
  captive, 43
  chest-beating of, 45–46, 48
  family groups, 46
  "John Daniel," 49
  jungle habitats, 43–45
  King Kong, 42, 43
  silverbacks, 46–47
  "Toto," 49
  warning routine of, 45–47

Hawks, 105–106; *Illus. 106*
Heyerdahl, Thor, 92
Hoyt, Maria, 49
Hugo, Victor, 98

Kon-Tiki, 92

Leopards, 24
Leopes, 6
Lions, 25–29; *Illus. 26, 28*
  behavior in zoos, 18

diminishing numbers of, 28–29
"Elsa," 25
hunting methods of, 25
man-eating, 27–28
show animals, 25
"Little Red Riding Hood," 5
Lycanthropy, 6

Octopus, 98–101; *Illus. 99*
Owls, 103–105; *Illus. 104*

Reptile Institute, 87

Schaller, George, 40, 43–45
Scorpions, 117–118; *Illus. 118*
Sharks, 89–98; *Illus. 94, 95, 96*
  attacks by, 89–93, 97
  blue shark, 96
  gray nurse shark, 97
  great white ("man-eater")
    shark, 93
  hammerhead shark, 97
  mako shark, 96–97
  repellent for, 92–93
  skeleton of, 97–98
  whale shark, 94
  white shark, 95
Snakes, 62–79; *Illus. 64, 69, 71, 73, 77*
  anaconda, 76
  asp, 74
  boa constrictor, 76–78
  cobra, 72–75
  copperhead, 69, 70
  coral snake, 63
  hognose snake (*or* puff adder,
    spreading adder, blowing
    viper), 72, 79
  king snake, 78–79

mamba, 72
poisonous species, 63–64
python, 76
rat (*or* chicken) snake, 78
rattlesnakes, 65–67
snakebite treatment, 67–69
taipan, 72
tiger snake, 72
venom of, 75
water moccasin, 70–71
Spiders, 110–117; *Illus. 111, 114*
black widow, 113–114
brown recluse, 115–117
tarantulas, 110–113
venom of, 113
wolf spiders, 112
Squid, giant, 101
Stoker, Bram, 52

Tigers, 18–24; *Illus. 20, 21*
behavior in zoo, 18–19
diminishing numbers of, 29
food requirements of, 22
hunting methods of, 20
Indian tigers, 22
Manchurian tiger, 22
man-eating, 19, 23–24
sizes of, 21–22
tame tigers, 22
teeth of, 20
Toilers of the Sea, 98
Tyrannosaurus rex, 1

Venomous animals, 118–119
Vultures, 107–109; *Illus. 108*

Wild dogs, 9
Wolves, 5–17; *Illus. 7, 9, 12, 14*
dire (*or* cave) wolf, 6–7
family life of, 13–14
hunting methods of, 20
"lone" wolf, 15
"prairie" wolf, 17–18
timber wolf, 7–12
werewolves, 5–6
wolf packs, 14–15